ENERGY, ENTROPY, AND THE FLOW OF NATURE

ENERGY, ENTROPY, AND THE FLOW OF NATURE

Thomas Fairchild Sherman

OXFORD
UNIVERSITY PRESS

OXFORD
UNIVERSITY PRESS

Oxford University Press is a department of the University of Oxford. It furthers the University's objective of excellence in research, scholarship, and education by publishing worldwide. Oxford is a registered trade mark of Oxford University Press in the UK and certain other countries.

Published in the United States of America by Oxford University Press
198 Madison Avenue, New York, NY 10016, United States of America.

Library of Congress Cataloging-in-Publication Data
Names: Sherman, Thomas Fairchild, author.
Title: Energy, entropy, and the flow of nature / Thomas F. Sherman.
Description: New York, NY : Oxford University Press, 2018. |
Includes bibliographical references and index.
Identifiers: LCCN 2017044266 | ISBN 9780190695354
Subjects: LCSH: Force and energy. | Entropy.
Classification: LCC QC73 .S534 2018 | DDC 531/.6—dc23
LC record available at https://lccn.loc.gov/2017044266

1 3 5 7 9 8 6 4 2

Printed by Sheridan Books, Inc., United States of America

For Kátia
Whose reverence and love are constants
In our world of endless change

CONTENTS

CONTENTS

ACKNOWLEDGMENTS

Whatever may be muddled or imprecise in this little book, I can truly claim as my own. All else I owe to others. A few of the giants who have developed our understanding of the laws of nature's flow are noted in the chapters of this book. Many others who contributed mightily have been neglected either by history itself or by my own wish to keep this book concise.

A few of the teachers and authors who stimulated my interest in the laws of nature are mentioned in the Introduction, but hundreds or indirectly hundreds of thousands of others go unmentioned and to them I should silently bow. In contemplating my own education, I wonder what, if anything, I would have uncovered by myself without the help of someone else. Would I know that the angles of a plane triangle must sum to two right angles had not Euclid demonstrated it over two thousand years ago?—and had not his proof been handed down through human generations and conveyed to me by a long-forgotten textbook and by Martha Neighbor, one of the many fine mathematics teachers at the Ithaca High School? I doubt that geometry would have competed successfully for my youthful energies against the attractions of outdoor sports had I not been blessed with such teachers as Ms. Neighbor. And so throughout my life inspired teachers and writers have awakened interests that otherwise would have remained dormant. I wish that all could be acknowledged. A few are to be found in the bibliography of this book.

A special and unusual acknowledgment is also in order. Sixteen years ago, had I lived in any other time but ours, my life would have ended in coronary artery disease. For four months in 2001, I was kept going by an implanted left ventricular assist device, by the constant care, love, and skill of nurses and doctors at the Maine Medical Center in Portland and the Brigham and Women's Hospital in Boston, and by the extraordinary devotion of my wife, Kátia, who brought cheerful companionship to me daily and converted my hospital room into something like a vacation cottage. Ultimately an unknown grieving family lovingly bestowed upon a stranger the living heart of their loved one, to accompany me for the remainder of my days. But for that blessed family, and for the surgical miracles of teams led by Dr. John Braxton in Maine and Dr. Gregory Couper in Boston, this book could not have been written.

In less dramatic but nevertheless very important ways, I am indebted to the Oxford University Press for the expert and kindly help of science editors Jeremy Lewis and Anna Langley and production editor Richa Jobin, and also to three anonymous reviewers for their wise perspectives and suggestions. To my friend and former colleague, Dr. Edward J. Kormondy, go my thanks for helpful comments on early chapters of the book. And at all times Kátia has been a constant source of sustainment. I thank her especially for taking time from her own scholarly writings on Cervantes to transform my hand-drawn diagrams into digital form.

Introduction

The central region of New York State is drained by numerous streams that tumble off the hills into the long, deep valleys of the Finger Lakes, carved in glacial times by the southerly advance of northern continental ice. As a child, I lived in a house perched close to the edge of one of these streams, Fall Creek, where it drops over a series of cascades on its way to Cayuga Lake. Our street ended above the last and largest of the falls, Ithaca Falls, and after a heavy spring rain I could hear through my bedroom window the roar of rampaging water. On many a summer evening, we would walk down to a look-out point above the gorge and gaze at the ever-changing curtains of plunging water as the sun sank behind the western hill and a hermit thrush trilled its evensong from a nearby wood. And, on many a summer afternoon, when the currents were not too swift, I swam at the foot of a small cascade in the deep fern- and hemlock-hung gorge above Ithaca Falls. No doubt it is memories of those childhood joys that make brooks and waterfalls such magical places for me, shaping the metaphors of my understanding and bringing to mind the ever-present creative energy of our world.

My life-long interest in nature and her energy began in the beauty of the Finger Lakes, but it matured at Oberlin College in Ohio, and in summers of guiding canoe trips in Algonquin Park and in the Superior–Quetico country, and in walking and bicycling through the British countryside as a graduate student at Oxford. At Oberlin were many gifted teachers in both the humanities and the sciences, among them George T. Scott, who lectured on the physical and chemical principles of biology in the choir loft of the old Second Church in Oberlin, with a sparkling zeal in his blue eyes that gave to the ions of sodium and potassium, to membrane potentials and osmotic pressure, a kind of divinity appropriate to the old church setting. From Luke E. Steiner and J. Arthur Campbell we received, in the old Severance Chemical Laboratory, an introduction to chemistry that was firmly rooted in its history, in the classical but quantitatively exact studies of Dalton, Faraday, Gay-Lussac, and Avogadro. I would not trade that experience for any modern course full of quantum mechanics and spectral analysis that did not explain where our ideas have come from. Much of the meaning of scientific ideas is revealed in their historical development, as the German philosopher-physicist Ernst Mach pointed out long ago.

The subject of thermodynamics developed in the 19th century from studies of the quantitative relationships between heat and mechanical motion and power, especially as evidenced in the working of the steam engine. As the concept of work expanded from the engineering realm of lifting weights or compressing springs to include the worlds of electrical, chemical, and biological phenomena, including physical transport and chemical synthesis, thermodynamics evolved into the broader subject of energetics, which deals with energy in all its forms.

Physical chemistry was taught at Oberlin by Art Campbell, a genial but commanding figure, tall and upright in stature, precise in diction, with the utmost clarity of thought—about, as I imagined, George Washington must have been. Campbell surprised us when he introduced thermodynamics with a cautionary, almost apologetic note—that this was a subject no one could understand the first time through. Coming from him—the professor who could make even the muddiest of waters clear—it was an unexpected comment. But he was right, at least for me: I did not understand it the first time, nor the second, even with the help of fine textbooks by Walter Moore and by Daniels and Alberty. The murkiness lay mostly beneath the unfathomable depths of *entropy*, a strange word with an enigmatic meaning that has been confusing humanity ever since the great Rudolf Clausius went to his Greek dictionary to coin the word in 1865. It might have been more merciful had he stuck to German. We will have lots to say about entropy later.

At Oxford, I had the great privilege to work with A. G. Ogston, as a physiology tutee at Balliol College and as a predoctoral research student in the Department of Biochemistry, which was headed at that time by Hans Krebs. Sandy Ogston was a physical biochemist of the first order who helped to construct in Oxford the world's second analytical ultracentrifuge, modeled after that built by Svedberg in Sweden. His students adored and revered him for his personal warmth and generosity and for his amazing ability to see the fundamental principles within complex problems. Two of his students, Baruch Blumberg and Oliver Smithies, went on to win Nobel Prizes.

Before I left for study in Oxford, my father, James Morgan Sherman, gave me several of his favorite books. One of them was William Mansfield Clark's *Topics in Physical Chemistry*. My father knew Clark, as they were colleagues in Lore Rogers's bacteriological research laboratory at the Department of Agriculture in Washington, D.C. Clark's book is a goldmine of insights into physical biochemistry. From Clark, I learned that free energy (the energy that can do work) is the product of an intensive factor (its potential or gradient) and an extensive factor (the capacity of the system), an idea that Clark himself may have gleaned from

J. N. Brönsted's *Physical Chemistry* (translated from the Danish by R. P. Bell, who also was at Balliol with A. G. Ogston). For all I know, Brönsted in turn may have been influenced by the writings of Ernst Mach. Much later, Brönsted's *Principles and Problems in Energetics*, his *Physical Chemistry*, and Ernst Mach's essays would all be important to my understanding of energetics.

When I went back to Oberlin to teach in 1966, my first assignment was to survey the essential principles of biochemistry for more than 300 introductory biology students assembled in Hall Auditorium. In 1953, I had listened to biology lectures in that vast auditorium; it was more than a little daunting to think that I would now have to give them. I decided to start from what seemed to me the most fundamental principles behind all living activity: the laws of energy. I planned those lectures with both care and trepidation, writing them out fully in a little study behind a cottage on Lake Mascoma in New Hampshire, where I was summering with my family. It was there by the waters that I saw how I wanted to approach the subject, how the principles could be most simply explained. I do not know quite where the ideas came from, but I suspect they owed a great deal to Clark and, through him, to Brönsted.

To my astonishment and great relief, the lectures were well received by the students, with one remarking that he had studied thermodynamics in chemistry the year before but only now saw what it meant. In the half-century since, I have pondered the subject over and over again and read or consulted scores of books on thermodynamics or energetics, but the essential viewpoint has remained the same: that all natural change involves a flow of something across a potential, across a gradient of some kind — like the water flowing through Fall Creek gorge near my childhood home in the Finger Lakes.

This book begins with an attempt to understand the historical origins of the concept of energy. What is energy, and where did the idea come from? This is a complicated brew, and it may not be everyone's cup of tea. Human understanding is always incomplete, but it is especially so if we fail to recognize the possible alternatives to our present views. Many of these alternatives were confronted and discussed by those who came before us, and to listen in on their conversations is often to see a light go on in our own understanding. Some readers may wish to skim or skip this historical introduction and go directly to the central core of the book, which begins with Chapter 3, but I hope that eventually their curiosity will bring them back to the beginnings. In Socrates's last days, as related by Plato, Socrates asked his friend, Cebes, to consider whether in our world of changing appearances there are entities that remain constant and invariable. Socrates was seeking intellectual support for his belief in the immortality of the human soul, but the development of natural philosophy— what we call science today—has sought other invariable realities that endure in all change. One of the most fundamental of these has come to be called *energy*.

Chapter 1

Origins of the Idea of Energy and Its Conservation

Our world is full of activity and change. The morning gives way to evening, the garden and the children grow, the school bus rumbles down the street, and the wind blows through the birch outside my study window. The river flows onward to the northern sea.

Since the time of the Ionian philosophers of ancient Greece, and long before then no doubt, people have sought to understand the origins of activity and change. The logical mind stumbles at the very question: How is change possible? How can A, which is not B, become B, which is not A? Are there rules or laws that govern what happens, so that some kind of constancy survives throughout the change? Many of the words used in describing nature (rules, laws, govern) derive from human social interactions, politics, and religion. The pre-Socratic philosophers of Ionia conceived of a unified nature, a universe operating under a common set of principles, at a time not far distant from when the Hebrews developed a monotheistic religion. One God, one nature, one set of moral and natural laws. If one believes in God—as Galileo, Kepler, Newton, Descartes, Leibniz, and most of the great founders of modern science did—then the natural laws are God's laws, and the discussion of what they are may invoke theological as well as philosophical ideas and feelings.

The study of energetics (or thermodynamics) deals with the most general principles of activity and change. Activity is a broader term than change for all change involves activity, but not all activity leads to change. This is evident in both human life and nature. A room full of political "activists" can work up a fury of activity, but only when they march together down the street is something likely to change—just as when the random Brownian activity of microscopic particles becomes the actual diffusive transport of materials from one place to another.

Energy is a measure of actual or potential activity. *Free energy* (a term we owe to Hermann Helmholtz) is a subset of energy—the energy (or activity) that can lead to change, or, to put it another way, the energy that can do work. *Bound energy* (another term we have from Helmholtz) is energy (or activity) that cannot—within the confines of the system we are considering—lead to change (or do work). Bound energy is the room full of activists without any outlet for their heat. The great William Rankine, for whom the Rankine temperature scale is named, got it wrong. He defined energy as the capacity to do work, and many subsequent books have set the stage for confusion by repeating this view. Only free energy can do work; bound energy cannot. *Total energy* equals free energy plus bound energy.

And what really is energy? Like many other fundamental elements of natural science—time, mass, force, space, or life—it eludes attempts to pin it down with a few words. We cannot hold it in our hands. The term did not enter the vocabulary of science until the 19th century, and revolutions in astronomy, physics, physiology, and chemistry were unleashed without Copernicus, Galileo, Kepler, Newton (even Newton!), Harvey, Lavoisier, or Dalton recognizing or using the concept of energy. The best we can do is to say that energy is a measure, indeed *the* most fundamental measure, of natural activity in all its multitude of forms.

Quantitative measurements in chemistry and physics gave rise, in the period 1775–1850, to two laws of conservation that govern all changes in nature: the law of the conservation of matter and that of the conservation of energy. The first of these laws was much more easily arrived at than the second. It depended only upon careful and accurate weighing (weight being one of the oldest of measurements), the recognition that "airs" or gases are forms of matter, and the invention of means for transferring and measuring gases in a quantitative manner. Antoine Lavoisier (1743–94) was the exemplar in putting these techniques and ideas together, keeping careful accounts of the weights of all substances, including gases, which are used up or are produced in chemical reactions. He found that weight in equals weight out, not only for total matter, but also for individual elements such as oxygen or mercury. Although every individual analysis was local, the results could be confirmed at other places and other times. Matter can be seen and touched. We can hardly imagine a world in which it should be created from nothing or disappear into nothing.

Energy, or the activity of matter, is something else. We cannot hold it in our hands nor easily define what it is we are talking about. The law of the conservation of energy, which amounted to a definition of energy itself, had a long historical development but crystallized in the 1840s in the thoughts and experiments of Julius Robert Mayer (1814–78), James Prescott Joule (1818–89), and Hermann Helmholtz (1821–94). These three very young men (all were in their mid-twenties at the time of their first publications on energy conservation) were heavily in debt to Lavoisier and the prior establishment of the conservation of matter. Lavoisier, Mayer, and Helmholtz were physiologists as well as chemists and physicists, interested in how oxidation, heat, and work are related in the animal body.

Mayer and Helmholtz received a medical education, and both became doctors (though Helmholtz gave up practice for academic life after 5 years).

RELIGIOUS AND PHILOSOPHICAL ORIGINS

The concept of energy and its conservation arose from many sources. One of these was deeply rooted in religion and metaphysics. Science is often portrayed as having an isolated life of its own in which cold reason and unbiased gathering of observations are uninfluenced by the hopes and fears of human existence. Religion and theology have been seen as adversaries to the advance of science, as in Andrew Dickson White's massive treatise on *A History of the Warfare of Science with Theology*. The burning of Giordano Bruno and the trial of Galileo by the Roman church, the burning of Michael Servetus by the Calvinists in Geneva, and the Scopes trial in Tennessee are well-known examples of the interference of religion with scientific thought. But most of the great contributors to the development of modern science (at least until the middle of the 19th century) held deeply religious views of nature as God's creation and of the study of nature as a way to understand the mind of God and to feel God's presence. Galileo, Kepler, Newton, Leibniz, Faraday, and many others were passionate in their belief that in pursuing science they were communing with the work and mind of God. Above the doorway of the 19th-century physics building on the campus of Bowdoin College today is the inscription, "The laws of nature are ideas in the mind of God."

If God created the world, and all these men thought he had, then surely he had done a good job of it. Leibniz seemed to think that it was not only good but perfect—a view that drew ridicule from Voltaire, who satirized Leibniz as Dr. Pangloss in *Candide*. Views of the goodness of the creation carried over into ideas on the physics of motion. The prime mover had acted at the creation, and the quantity of motion in the world must always be the same, or so thought both Descartes and Leibniz. Newton was not so sure as he saw evidence that motion gets run down by viscous forces (anticipating, if you like, the second law of thermodynamics), so that God, intimately involved in the world, might need occasionally to intervene to restore lost motion. Leibniz and Newton quarreled, not only about who first invented the calculus, but about who had the truer view of God. Science and religion were not separate elements in the thinking of Leibniz or Newton, and the idea of the conservation of motion was as much religious as scientific. As James Joule was to put it in 1843, as he made his measurements of the mechanical equivalence of heat: "I shall lose no time in repeating and extending these experiments, being satisfied that the grand agents of nature are by the Creator's fiat indestructible; and that, wherever mechanical force is expended, an exact equivalent of heat is always obtained."

PHYSICAL MOTION AS A MODEL FOR CHANGE

The model for all change in nature was motion itself, and as science and philosophy adopted an increasingly mechanistic view of the world in the 17th to 19th centuries (a view encouraged by increasing experience with mechanical devices such as waterwheels, windmills, pumps, and clocks), all change came to be viewed as a kind of motion. So, to dispute, as followers of Leibniz and Newton did, whether the quantity of motion was always the same was to probe the deeper question of whether the quantity of all activity and change remained invariant—whether behind all the change was something (other than God) that did not change. This required, paradoxically, a measure of change that could be shown not to change.

The measure of motion was itself controversial. If a body is at rest, it accelerates (acquires increasing motion) as a force is impressed upon it, obeying Newton's second law, $f = ma$ (or more strictly, as Newton put it, $f = \Delta(mv)$). The quantity of motion in a body is, according to Newton, dependent upon its mass as well as its velocity and is called its *momentum, mv*. The velocity of the body, and hence also its momentum, increases linearly over the time that the force acts upon the body. In free fall, a constant (almost) gravitational force provides a uniform acceleration, g, so that a velocity, $v = 0$ at the start becomes for a subsequent time, t:

$$v = gt$$

At the end of the fall, the time is t_m and the velocity is at its maximum, v_m:

$$v_m = gt_m \quad \text{or} \quad t_m = \frac{v_m}{g} \tag{1.1}$$

Because the acceleration has been uniform, the average velocity for the fall is:

$$\bar{v} = \frac{0 + v_m}{2} = \frac{1}{2}v_m$$

The total height, h_m, of the fall is:

$$h_m = \bar{v}t_m = \frac{1}{2}v_m t_m \quad \text{or} \quad t_m = \frac{2h_m}{v_m} \tag{1.2}$$

Equating the values for t_m from Equations (1.1) and (1.2) above:

$$\frac{v_m}{g} = \frac{2h_m}{v_m} \quad \text{or} \quad v_m^2 = 2h_m g \tag{1.3}$$

Let us drop the subscript, m, realizing that any point during the fall could be regarded as the end point of a fall, so that the equations describe relations among v, t, h, and g for all points of a fall. Equation 1.1 says that velocity is proportional to time, a relationship discovered by Galileo. But Equation 1.3 says that velocity squared is proportional to distance, which Galileo failed to determine, but which was discovered later in the 17th century. Ernst Mach (for whom the dimensionless Mach number—the ratio of speed to the speed of sound—is named), a physicist-philosopher and historian of science, speculated that if Galileo had discovered Equation 1.3 instead of Equation 1.1, the subsequent course of physics might have been very different. Equation 1.1 is related to forces, while Equation 1.3 is related most directly to what we now call energy. Put an m (for mass) in front of the v of Equation 1.1, and we have Newton's momentum, mv. Put an m in front of the v^2 of Equation 1.3, and it becomes mv^2, which Leibniz called "*vis viva*" ("living force"—a force associated not with life but with motion), and which today, with a ½ in front of it, is called kinetic energy. With the introduction of mass to each side, Equation 1.3 becomes:

$$\frac{1}{2}mv^2 = mgh$$

Each term of this equation has the dimensions of energy, $ML^{-2}T^2$. Today, this equation is taken to be an expression of a conservation of mechanical energy, in which the process of falling converts the potential energy of the weight at height (mgh) into the kinetic energy of the mass in motion $\left(\frac{1}{2}mv^2\right)$.

For Newton and Descartes, momentum was the proper measure for a quantity of motion, but for Leibniz the correct measure was *vis viva*, and a great deal of controversy ensued between followers of the one or the other. Either view has its justification. Any body in motion has, of course, both momentum and *vis viva* (kinetic energy)—the first represents the product of a force acting upon a mass through time, while the second represents the product of a force acting upon a mass through distance. If momentum is viewed, like velocity, as a vectorial quantity (having direction as well as amplitude), then its conservation follows directly from Newton's second and third laws. By the second law, forces cause (over time) changes in momentum, and, by the third law, every force has an equal and opposite force. Within an isolated system, a force on one part is coupled to an opposite force on another part, and since a common duration of time applies to these forces, a change in momentum of one part is accompanied by an equal and opposite change in momentum of another, and hence the total momentum of the system must remain constant. Hence, the second and third laws lead directly to the principle of the conservation of momentum—and if momentum is the true measure of the quantity of motion, then the quantity of motion must always remain the same. Yet two equal masses moving toward one another and sticking to one another in collision completely annihilate each other's motion. This is a strange view of the conservation of motion—even if momentum, due to its vectorial definition (invented by Christiaan Huygens), has been conserved.

While Newton's laws lead to the conservation of momentum, mv, they do not give us the conservation of kinetic energy, *vis viva, mv²*. Whereas opposing forces must share a common duration of time, they do not necessarily share a common change in distance. One has only to think of a rock falling to Earth: the rock moves a great distance affected by the force from the Earth, but the massive Earth moves hardly at all in response to the force from the rock, and, in the moments of impact, there is no way to measure what the distances are. *Vis viva* (kinetic energy) is created in free fall and hence is not conserved, and it disappears (in the absence of an elastic bounce) when the object hits the ground. To save the conservation of causes and effects in the activity of free fall, one must suppose two further factors: a conversion of potential motion into actual motion (potential energy into kinetic energy) during the free fall and a conversion of motion into something else (thermal energy) at the end of the fall.

The impact of falling objects was studied experimentally by Willems Gravesande (1722), who dropped identical balls from varying heights into soft clay and measured the depth of the impression created by the ball in the clay. A ball dropped from 4 feet achieves twice the velocity at impact as one dropped from 1 foot (velocity in free fall varies with the square root of the distance). Gravesande found that twice the velocity gave four times the depth of impression, so the effect of the ball depended on velocity squared rather than velocity—on *vis viva* rather than momentum. Note also that the effect on the clay varied linearly with the height of the drop.

The difference between momentum and kinetic energy is dramatically illustrated in the firing of a gun. A 22-calibre rifle weighs about 3 kilograms while its bullet weighs 3 grams. The exploding gases from the cartridge push equally (almost) on the rifle and the bullet, so that the rifle and bullet achieve the same momentum in opposite directions. But since the rifle is 1,000 times heavier than the bullet, the bullet travels 1,000 times faster than the recoiling rifle. The rifle and its bullet have the same mv, but the mv^2 of the bullet is 1,000 times greater than that of the rifle—even though they have been subject to the same (but opposite) forces in firing. The potential effect of the recoiling rifle is nothing compared to that of the speeding bullet.

The squared velocity of *vis viva* makes it always a positive term with respect to any direction—which is to say that it is a scalar, not a vectorial quantity. If two masses in collision bounce off one another in a perfectly elastic manner, like an ideal ball bouncing to the same height from which it was dropped, then *vis viva* is conserved, but otherwise some of it is lost. Leibniz proposed that losses of *vis viva* in impacts were due to the transfer of motion from macroscopic bodies to their tiny constituent particles—so that total *vis viva* was conserved. This sounds prescient of the modern view of mechanical energy being converted to thermal energy, so that total energy is conserved.

CONSERVATION SUGGESTED BY SIMPLE MACHINES

A suggestion of conservation came from studies of the inclined plane, the pendulum, and the lever. Galileo noted that a ball rolling down an inclined plane achieved a velocity just sufficient (almost) for it to roll up another inclined plane to the same height from which it had started. The second plane could have a different incline from the first—the rolling ball still sought the level from which it had started. Similarly, in the swinging pendulum, the bob sought the same height from which it had started, even with changes in the amplitude of its swing. With either inclined planes or the pendulum, the mass starts with zero velocity, gains velocity as it descends to its lowest point, and loses velocity as it ascends back to the same height from which it started. It is as if height can be traded for velocity (actually velocity squared) or velocity for height, with the rate of exchange always the same—as if it were agreed in the banking system of nature that the one is as good as the other because they are really in some way the same thing. So, too, in a balanced lever is a conservation of a different sort, where the product of a weight times the distance it moves up equals, at the other end of the lever, the product of weight and the distance it moves down. Change the weights and where they sit on the lever arm: if the lever is initially balanced, then, with a slight movement, what is gained in work (weight times height) at one end is exactly lost at the other.

MACHINES CANNOT MAINTAIN PERPETUAL MOTION BY THEMSELVES

A conservation of activity (energy) implies that it cannot be created from nothing nor disappear into nothing. Newton, as we have said, had concerns about the latter and wondered if God might not have to restore his world from time to time. But almost every natural philosopher agreed that activity could not be created without some prior cause. This idea is often referred to as the *impossibility of perpetual motion*, or the impossibility of getting motive power from nothing. If a waterwheel is built to run a gristmill, it needs an external source of flowing water. It cannot be capable of pumping water back upstream in sufficient quantity to maintain its own water supply while grinding the wheat to flour. No machine can keep running, doing work in the face of friction, without a source of power. Can anyone prove this from Newton's laws? No, and many tried to disprove it by inventing a perpetual motion machine, but all attempts failed. Humanity, and nature itself, seemed constrained by this limitation—but the limitation itself implied one-half at least of a conservation principle.

NATURE REVEALED AS A PLENUM OF INTERACTING, RECIPROCAL FORCES

At the turn of the 19th century, nature began to reveal new forces, new relationships, symmetries, and reciprocities that had been hidden until that time, and that, one after another, came as complete surprises. Much can be traced to a helpless frog and to a deep-thinking philosopher who was also a physicist.

The philosopher-physicist was Immanuel Kant (1724–1804). Kant applied physical dynamics to the deep history of the universe, theorizing on the nature of galaxies as clusters of stars and on the origin of our solar system, and it was he who first suggested that the tides act as a frictional brake on the rotation of the Earth, slowing it down so that earthly days have grown longer over the course of the Earth's history. And in his philosophy, Kant saw an order in nature that was reflected in the human mind, one in which all parts were related to the whole. For Kant, as for Aristotle and Descartes, the world was a continuous plenum (not one of discrete atoms and the void), filled with forces of attraction and repulsion, and the unity of the world implied that all the various forces—gravitational, electric, magnetic, and the like—could be converted one into another. Kant's view of nature influenced Johann Ritter, who tried in 1798 to create a theory of electrochemistry that melded electrical and chemical forces. In 1798 also, Samuel Taylor Coleridge—who was a natural philosopher as well as a poet and friend of William Wordsworth—spent a year in Germany and became an enthusiast of Kant's philosophy. Coleridge brought Kant's ideas back to England, where they appealed to Humphrey Davy (1778–1829) and his protégé Michael Faraday (1791–1867) at the Royal Institution and to others who wanted a less materialistic and more organic view of nature than that provided by mechanistic philosophers such as Locke and Hume. On the continent, Kant's views heavily influenced Robert Mayer and Hermann Helmholtz, both of whom were deeply philosophical in their thinking about physical and biological nature.

The frog that launched a new era in science was in the laboratory of Luigi Galvani (1737–98). At the end of the 18th century, electricity was known only in its static form, created by friction and stored on capacitors such as Leiden jars. But, in 1791, Galvani noticed that the leg muscles of a dissected frog sometimes twitched when a sciatic nerve was touched with a metal scalpel held by someone close to the discharge of a static-electrical generator. The sparking generator was analogous to lightning, which Benjamin Franklin had already shown to be electrical. So Galvani did one of those improbable experiments in the history of science (even more outrageous than Franklin's kite-flying): he hung freshly dissected nerve–muscle preparations on an iron railing outside his laboratory, secured by brass hooks, to see whether atmospheric electricity would cause them to twitch. He did not have to wait for lightning. At the moment the preparations were placed on the railing, some of them twitched. This result was entirely unexpected. As Alessandro Volta (1745–1827) showed in repeating Galvani's experiment, it only worked when dissimilar metals (e.g., brass and iron or copper and zinc) were used. (Today, we can say that copper takes electrons from zinc, or iron from copper, so that two metals together create a pair of negative and positive electrodes.) Galvani was not put off by Volta's criticisms, for he had also noticed that the nerve–muscle preparations responded to injured tissue of the same or a different frog, and he became convinced that the frogs themselves were sources of electricity to which their own nerves were exceedingly sensitive. This "animal electricity" (which Volta initially dismissed as nonsense) created great excitement for both physiologists and physicists, opening up an entirely new field of electrophysiology. Among the early pioneers in this new field was the young Hermann Helmholtz, who succeeded (in 1850) in determining that the velocity of stimulus conduction in the sciatic nerve of a frog is about 30 meters/sec. Every electrocardiogram done today has its historical roots in Galvani's frogs.

But an even greater contribution to our story of energy lay in the brass hooks and iron railing to which the frog nerves reacted. Volta created from dissimilar metals the first Voltaic piles—stacks of copper and zinc, for example, each metal separated from the next by paper soaked in salt water. These batteries were the first source of current electricity—a steady flow instead of a momentary discharge of electricity—and they would lead to the discovery that a multitude of forces in nature (electrical, magnetic, chemical, mechanical, thermal, and radiant) are all intertwined in reciprocal interactions with one another. A great many investigators were involved in the physical-chemical revolution of the next half-century, but none was so central as Michael Faraday.

Faraday's story is one of most moving in the entire history of science. The son of a blacksmith, he was born into the London poverty that would soon be documented by Charles Dickens. He lived with his poor but devoutly religious family in a small flat over a coach house and received only a smattering of elementary education before he was put to work to help support his family at age 13; he was apprenticed the next year to a bookbinder. But an irrepressible curiosity lay within his youthful energies, and he began to read the books that he bound and others in the bookshop, including the article on electricity in the *Encyclopedia Brittanica* and Jane Marcet's *Conversations on Chemistry*, a book inspired by Mrs.

Marcet's attendance at Humphrey Davy's lectures at the Royal Institution. He made his own voltaic pile out of seven copper half-pennies, seven disks of zinc he cut from a sheet of zinc, and paper disks soaked in salt water. At age 21, Faraday was desperate to escape the world of bookbinding for the world of chemistry and electricity. He was given tickets to attend four of Humphrey Davy's lectures, took copious notes, and bound the notes into a book that he sent to Davy with a letter begging for any kind of job that Davy could give him at the Royal Institution. Davy, one of the world's foremost chemists, took Faraday on as an assistant. Like Faraday, Davy had no formal education in science but, as a youth, developed a passion for chemistry. Although but 13 years older than Faraday, he had by 1812 become England's most renowned chemist. He was also something of a poet, a friend of Coleridge, and, through Coleridge, of Wordsworth. At age 22, he was asked by Coleridge and Wordsworth to edit their *Lyrical Ballads*. He made many great discoveries in chemistry, including the isolation of pure sodium and potassium and proof of the elemental nature of chlorine, but as someone later said, his greatest discovery may have been Michael Faraday.

When Faraday joined Davy at the Royal Institution, Davy was at the forefront of the new field of electrochemistry, which had grown out of Volta's invention of the battery. Volta thought that his pile involved only a displacement of electrical charges at the surfaces of the dissimilar metals and that the pile would give an unlimited movement of electricity (a perpetual motion, in effect), but it soon became apparent that chemical reactions were going on within the pile. In 1800, William Nicholson and Anthony Carlisle discovered that when the wires from a pile were immersed in water, hydrogen was produced at one wire and oxygen at the other. Here was the most common chemical reagent, water, being decomposed by electricity. As chemical reactions gave rise to the electricity produced in the pile, so the electricity of the pile could give rise to further chemical reactions. Davy used the electricity of a massive pile to separate the pure alkaline metals of sodium and potassium from their oxides. The natural forces of chemistry and electricity were interrelated, as Kant had foreseen.

Furthermore, when chemistry and electricity interact, they do so in definite proportions. Faraday, in 1834, reported what are today called Faraday's laws of electrolysis: (1) the quantity of a chemical produced at an electrode is proportional to the quantity of electricity that passes through the electrode; and (2) the chemicals produced at the two electrodes are different, but their quantities, expressed in chemical equivalents, are the same. In effect, Faraday discovered that, in the interactions of electricity and chemistry, a definite rate of exchange is upheld by nature. This is analogous to the conservation of matter recorded in Lavoisier's balance sheets, but it goes somewhat beyond it, into the realm of chemical and electrical forces or activities. Chemical reactions can give rise to a quantity of electrical displacement and electrical flow over time to a quantity of chemical reaction, but always tit-for-tat—cause is equivalent to its effect.

In 1820, news arrived at the Royal Institution of Hans Christian Oersted's surprising discovery that an electric current deflects a magnetic compass needle. The magnetic poles of the compass align themselves not parallel to the current but perpendicular to it, the north pole pointing one way when the compass lies beneath the electrical wire and the opposite way when it lies above it. The magnetic force created by the current, Oersted noted, is wrapped circularly around the wire. This was not like the straight-line forces of gravity or static electricity, but something entirely new.

The following year, 1821, Faraday surmised that the actions of electrical currents and magnets might be reciprocal, just as are the actions of electrical currents and chemical reactions. If the electrical current had a circular force around it that could make a magnetic needle rotate, so a magnet might make a current-carrying wire rotate. Faraday invented an apparatus to demonstrate this action, placing a magnet upright in a bowl of mercury, with only its end exposed. One end of a voltaic pile was connected to a loose wire that dipped into the mercury close to the magnet. A fixed wire connected the mercury to the other end of the pile. When the current was turned on, the loose wire began to rotate around the head of the magnet. Faraday's demonstration was a sensation. He had invented the first electric motor and shown the convertibility of electrical activity to mechanical activity. The wire and the magnet could change places in their circular dance: if the wire were fixed in place, Faraday could get a magnet to rotate around it.

Faraday's view of the reciprocity of natural forces suggested to him that since an electric current has a magnetic force around it, as Oersted had shown, then magnetic forces might induce an electric current. But simply placing a strong magnet next to a coil of wire (call it the *secondary coil*) will not induce a current in the coil—it could not, else one would have a source of perpetual motion, a motion created from no motion. The key, Faraday found in 1831, was that the wire had to be exposed to a *changing* magnetic field in order for a current to be induced in it. The changing magnetic field required an activity of some kind to cause the change. That activity could be the making and breaking of an electrical current in a primary coil serving as an electromagnet, or it could be the mechanical motion of an iron magnet near or through the secondary coil. By such means, mechanical activity could generate electrical activity through the intermediary of changing magnetic fields. Faraday had found the means to produce electrical activity from mechanical activity, just as he had found 10 years earlier the means for converting electrical activity into mechanical motion. The principles of both the electric generator and the electric motor were now at hand, derived ultimately from the frogs in Galvani's laboratory 40 years earlier.

By the 1840s, humanity had a much richer view of the various forces of nature, of its ways and means of expressing activity, of the reciprocity and symmetry of natural processes, than it had just a half-century before in the time of Lavoisier. In the 21st century, it may be hard to appreciate this change for the success of technology has created a world of electronic gadgets that is beyond any individual's ability to understand, and the idea that one kind of action can be transformed into something quite different is taken so much for granted that few stop to think about it. No longer do we exclaim, "What hath God wrought?" at the transmission of a message across a telegraph wire. Instead, the reaction is likely to be, "Why do we need a wire?" The enrichment of science in the period 1790–1840 created a need to find a common understanding, to seek a common measure, of all these natural processes.

MECHANICAL WORK AND HEAT

The first common measure would come through a comparison of mechanical work with heat. No one needs to be a natural philosopher to know that mechanical work often generates heat for physical activity makes our bodies warm, and the harder we exercise the hotter we get. Aristotle thought that our warm breath is a means for getting rid of excess warmth—hence we breathe harder when we exercise in order to get rid of the greater heat produced. Friction can produce great heat: an Aboriginal method for starting a fire is to rotate a dry stick very quickly upon another dry piece of wood. The wooden wheels of a coach had to be kept well-greased lest they char or even catch fire in rotating fast on a wooden axle. (Motorists descending Pike's Peak today are required to stop part way down to let their brakes cool off.) And Benjamin Thompson (Count Rumford, 1753–1814) famously demonstrated (in 1797) that an indefinite amount of heat is generated in the process of boring out the barrels of cannons and that heat cannot therefore be a material substance, but must be "a kind of motion."

A method for measuring heat was devised about 1760 by Joseph Black (1728–99). The method depended entirely on the prior invention of the thermometer, for heat was assessed by the temperature change it caused in a definite quantity of matter. Today, a standard unit of heat is the *calorie*, which is the amount of heat required to raise 1 gram of water by 1° centigrade (more exactly, from 14.5° to 15.5°C). Black discovered that much heat (which he called latent heat) is required to melt ice to water without any change of temperature. This finding was utilized by Lavoisier and Pierre Laplace (1749–1827), who invented an ice calorimeter for determining heat (caloric) by the amount of ice melted.

Mechanical work was measured as the height to which a stipulated weight could be lifted; that is, as the product of a force times a distance. But is there any way to join heat and mechanical work together in a common measure? Can anyone show that they must be dimensionally equivalent to one another? Enter our three main protagonists, Mayer, Joule, and Helmholtz.

ROBERT MAYER

In early 1840, 25-year-old Robert Mayer, a recent medical graduate from Heilbronn (Germany), obtained a position as ship's doctor on a voyage to the Dutch East Indies. The young man seems to have had a deep philosophical curiosity as well as a sense of adventure, and, as a child playing with little waterwheels in a brook near his home, he had learned the futility of trying to make a perpetual motion machine. He was interested in the physiology of the human body, which he tended to view in mechanical terms, but he had had very little formal training in physics. A year after leaving for the East Indies, he was back in Heilbronn informally studying physics while earning his living as a physician, and a year after that (1842) he published a ground-breaking paper on the conservation of nature's forces (energy) that contained the first estimation of the quantitative equivalence of mechanical work and heat. How Mayer went in 2 years from a fledgling doctor heading for the East Indies to a discoverer of the concept of energy conservation is one of the strangest stories in the entire history of science.

On the outward journey to the East Indies, the crew was healthy, and Mayer had little to do but read his physiology books, taking special interest in the chemistry of the blood. When they reached the East Indies in June, however, the sailors suffered an outbreak of respiratory infections. Mayer did what his medical education had taught him to do: he bled his patients from a vein in the arm. The practice of bloodletting was rooted in the ancient Greek theory of the four humors of the body and in the idea that when the body developed a fever, warm blood was in excess and should be drained away in order to restore balance of the humors.

The man who would soon formulate the conservation of energy was practicing ancient medicine that was in curious disjunction from the progress of physiological science. It was 212 years after William Harvey's *On the Motion of the Heart and Blood in Animals* (1628), which revealed the torrent of blood that circulates throughout the human body, and 107 years after Stephen Hales's

Haemastaticks (1733), which demonstrated the relation between blood pressure and blood flow. The maintenance of blood pressure is dependent on the maintenance of blood volume, but Mayer was not thinking here in mechanical terms. Apparently, Mayer's bloodletting was moderate, and all his shipmates recovered, but this crude medical practice may have contributed over the ages to the deaths of many patients, including the first president of the United States who, stricken with a respiratory ailment, was bled three times at Mount Vernon (the first time by himself, the second and third times by his doctors) before he died.

In performing this outdated medical procedure, however, Mayer made an unexpected observation: that the venous blood from his sick sailors seemed unusually red, almost as if he had hit an artery rather than a vein. It had recently been shown by Gustav Magnus that the redness of arterial blood is associated with high oxygen content—that arterial blood is higher in oxygen but lower in carbon dioxide than venous blood. Mayer also knew from his training in chemistry that combustion experiments, starting with those of Lavoisier and Laplace, had shown a proportionality between the amount of fuel oxidized (either in a flame or in the animal body) and the amounts of carbon dioxide and of warmth (heat, caloric) produced—so much oxygen consumed gives so much heat.

Mayer thought that in the tropics the human body would use less oxygen because it needed to produce less heat and that this could account for the venous blood appearing redder (having more oxygen). He was evidently thinking of the human body as a warm physical object that would cool less rapidly in a warm environment (the temperature gradient being smaller) and hence would produce less heat and use less oxygen in keeping itself warm. This is a reasonable view, and one that shows that Mayer was thinking like a physicist who viewed the living body as following the same laws as inorganic objects, an important point of view in a time when many thought that life made up its own rules and exhibited vital forces that were unknown in the nonliving world. It is a view that seems to have satisfied not only Mayer but those who have commented since on this strange bloodletting observation. The physiology, however, is more complex than Mayer realized, and his explanation was probably erroneous.

The human body has a zone of thermal comfort in which the naked body at rest uses the least amount of oxygen and produces the least amount of heat. This zone of thermal neutrality is at environmental temperatures of about $25 - 27°$ C $(77 - 80°$ F). At cooler temperatures, the body will produce more heat (as, for example, by shivering) in order to stay warm. But at warmer temperatures it will produce more heat (as by greater blood flow to the skin and by perspiration) in order to stay cool! We do not know the thermal condition of Mayer's patients in the tropics, but a likely explanation for the greater redness of the venous blood is that the blood flow to their arms was increased in an attempt by the body to cool itself. With blood flow to the arms increased, the drop in oxygen concentration between the arterial and venous blood would be less, and the venous blood would be redder. The venous blood in the arms under such circumstances was not representative of the average venous blood throughout the body.

In ruminating on his observation, however, it occurred to Mayer that the human body uses oxygen in the production of muscular work as well as for the generation of heat. It is well known that we breathe more deeply and frequently in exercise than at rest, and the harder our muscles work, the faster and more deeply we breathe. Work and heat are both related to oxygen in the human body, and since it was known that heat production bears a definite relationship to oxygen utilization, it seemed probable to Mayer that the mechanical work of the body does, too, and that if both heat and work have an exact equivalence to amount of oxidation, they must have an equivalence to one another.

This was not the first nor the last time in the history of science that from lowly beginnings, from dubious procedures and initial misinterpretations, great ideas have unfurled. Mayer saw the interrelations of oxygen, work, and heat in the animal body as emblematic of a unity that exists in all of nature, and, like many others before him, from Aristotle to William Harvey and beyond, he seems to have seized on the idea that nature does nothing in vain, that justice (if you like) runs through all nature, that in all processes a cause must be equal to its effect, that nothing fundamental is really lost as one thing changes to another. Mayer went as quickly as possible back to his native Heilbronn to study physics while he practiced medicine and to think deeply on the unity he saw in natural processes.

Mayer was searching for constancy in the midst of natural change. Just as Lavoisier had found a constancy of material weights in chemical transformations, so Mayer felt that there should be a constancy of forces, which he regarded as causes, in change. For consistency in nature, a cause must equal its effect. As an example of this, Mayer considered the case of a free-falling body that increases its *vis viva* (kinetic energy) in direct proportion to its loss of "fall force" (potential energy). Fall force can become *vis viva*, or, if an object is thrown upward, *vis viva* becomes fall force, so that in some fundamental way the one is equivalent to the other, and the sum of the two is constant. The *vis viva*, however, can often be destroyed by friction or collision, producing heat, and, if cause equals effect, the heat produced must be exactly equivalent to the *vis viva* that vanishes. Then Mayer hit upon a brilliant way to assess this equivalence, and, in 1842, he published a value for the mechanical equivalence of heat.

Mayer based his value for the mechanical equivalence of heat not on new experiments of his own, as Joule was undertaking (see next section), but upon a calculation based on data that already existed in the chemical literature. He used an ingenious argument, which he did not fully explain until 3 years later. Here is the explanation in contemporary terms:

A gas such as air can be heated at constant volume (in a closed container) so that, as its temperature increases, its volume cannot expand. Or it can be heated at constant pressure (in an open container) so that, as its temperature goes up, it also expands against the atmospheric pressure. The amount of heat required to raise the temperature of the gas by $1°$ is more when the gas is heated at constant pressure than at constant volume. The extra heat is required because work is done when the gas expands. This work may consist of two parts: internal work done on the gas itself, overcoming molecular attractions within the gas so that it can expand, and external work done by the gas in pushing back the atmosphere around it. In an ideal gas, molecular attractions do not exist, so that no internal work is necessary, and the only work is external work. The difference in heat required to raise the temperature of an ideal gas at constant pressure versus constant volume should therefore be equivalent to the external work that the gas does in expanding. If we can determine both the heat difference and the work, then we can calculate the work equivalence of heat. Mayer used data available for air and implicitly assumed that all the work of expansion was external, an assumption that Joule and Kelvin subsequently questioned. The latter's experiments to demonstrate internal work on expansion showed, however, that Mayer was very nearly justified in ignoring it for air.

We will use the ideal gas law and modern measurements to illustrate Mayer's method for calculating the work equivalence of heat, work being expressed in joules (named for James Joule) and heat in calories. One joule $= 10^7$ ergs, where 1 erg $= 1$ dyne-cm, a dyne being the force required to accelerate a mass of 1 gram by $1 \text{ cm}/\text{sec}^2$. (One joule also equals 1 newton-meter, where a newton is the force required to accelerate a mass of 1 kilogram by 1 meter/sec^2.) One calorie (with a lowercase "c") is the heat required to raise the temperature of 1 gram of water by 1° Centigrade. The kilocalorie=1000 calories is also called a Calorie (but with an uppercase "C"). It is kilocalories that are referred to in human nutrition. A reasonably active person may need 2,000 Calories per day to maintain weight. He/she would starve on 2,000 calories per day.

The ideal gas law states that $PV = nRT$, where P is pressure, V is volume, n is the number of moles of gas, R is the gas constant, and T is the absolute (Kelvin) temperature. Of the five factors, P, V, n, and T are variables, but R, as its name implies, is a constant. The number of moles (number of molecular weights in grams), as a measure of the amount of gas involved, is obviously important in the laboratory, but in theoretical discussions it is a nuisance, so let us work with 1 mole, so that the ideal gas law becomes $PV = (1)RT = RT$. The term on the left, PV, has the dimensions of energy, so the term on the right must also have the dimensions of energy (the dimension of moles being implicit in the unwritten (1). The gas constant, R, has the dimensions of energy per degree per mole.

The gas constant, R, is the derivative (change) of PV energy with temperature:

$$R = \frac{d(PV)}{dT} = P\frac{dV}{dT} + V\frac{dP}{dT}$$

When pressure is constant, $\dfrac{dP}{dT} = 0$, and the second term above vanishes, so that

$$R = P\frac{dV}{dT} \quad \text{or} \quad RdT = PdV$$

The heat added to a gas to raise its temperature by an infinitesimal amount dT at constant volume is denoted by $C_v dT$, where C_v is its heat capacity at constant volume. Likewise, the heat added to a gas at constant pressure is $C_p dT$, where C_p is its heat capacity at constant pressure. We have already seen that, for an ideal gas, $C_p dT$ is greater than $C_v dT$ by the amount of external work done, which is PdV or, as we saw earlier, RdT. Then:

$$C_p dT = C_v dT + RdT \quad \text{or} \quad C_p - C_v = R$$

From experimental data for the heat capacities, C_p and C_v, which were available to Mayer, the value of R can be found by difference, expressed in calories of heat per mole-degree: R = 1.987 calories/mole-degree. If 1 mole of an ideal gas is heated by 1° Kelvin (or Centigrade) at constant pressure, then 1.987 calories of the heat are expended to do external work. If we can calculate how much external work is done under these same circumstances, then we can determine the work equivalence of heat.

The external work done is that of the constant pressure of the heated gas pushing back the atmosphere through a volume equal to the expansion of the gas. The volume of expansion is the volume of the gas (22.4 liters for a mole of gas at $0°$C or $273.1°$K) times the fractional increase in volume for a rise of 1° Kelvin. The fractional increase is $1/273$, so that volume increase in the gas is $22.4/273 = .082$ liter $= 82$ cm.[3] The pressure of the standard atmosphere against which the gas expands is

1.01325 x 10^6 dynes / cm.^{2} Hence the work of expansion of the gas is: $P\Delta V = (1.01325 \cdot 10^6)(82) = 83.1 \cdot 10^6$ ergs $= 8.31$ joules . The equivalence between heat and work is then:

$$1.987 \text{ calories} = 8.31 \text{ joules}$$

$$1 \text{ calorie} = 8.31 / 1.987 = 4.182 \text{ joules}$$

Mayer expressed his result in terms of the height to which 1 gram of water could be lifted by the work equivalent to the heat required to raise its temperature $1°$. The heat required is 1.987 calories, and the work equivalent is 4.182 joules. How high can 4.182 joules of work lift 1 gram?

$$\text{work in joules} = (\text{force in newtons})(\text{height in meters})$$
$$(\text{force in newtons}) = (\text{mass in kilograms})$$
$$\times \; (\text{gravitational accel. in meters} / \sec^2)$$
$$= (.001)(9.8) = .098 \text{ newtons}$$
$$\text{height} = \text{work} \div \text{force} = 4.182 / .098 = 427 \text{ meters}$$

This result says that 1 gram of water could be raised to a height of 427 meters by the energy required to heat it by only $1°C$. Using the data available at his time, Mayer calculated a value of 367 meters. This was the first (1842) published estimate of the equivalence between mechanical work and heat, but Mayer did not explain his method until 1845, and few people even then understood what he was doing.

Mayer was only 28 years old when his result was published. Much of his time was occupied by his medical practice, but his independent study of physics and chemistry gave him perspectives that others had missed, a little like the young Einstein 60 years later. Who could suppose that the heat capacities of air measured under different circumstances in the laboratory would allow one to say: " 1° of heat = 1 gram at a height of 367 meters"? This one calculation did not prove that heat and work always have the same relation, but Mayer saw nature as a unity in which, just as material substance is conserved in chemical reactions, imponderable force (or energy as it was soon called) must be conserved as it passes from one physical form to another.

JAMES JOULE

At the very time that young Mayer in Germany was publishing his result for the mechanical equivalent of heat, the even younger James Joule in Salford, near Manchester in England, was close to announcing his own findings. Joule's father was a wealthy brewer who hired private tutors for his sons, including Manchester's illustrious chemist John Dalton, and who purchased the equipment necessary to set James up with his own physics laboratory.

The young Joule was interested in whether the new electric motors that had recently become available as the result of Faraday's discoveries might replace steam as motive power for the family's brewery. He soon found that the metals required to build the batteries to drive an electric motor were much more expensive than the coal required for a steam engine, but, in the process, he discovered that the power delivered by an electric motor was proportional both to the current, i, and the voltage, V, driving the motor. Georg Ohm (1787–1854) had previously found that current and voltage are proportional to each other through a third factor, the resistance, R. Summarizing these relations in algebraic terms (and assuming that proportionality coefficients can be made equal to unity by suitable choice of units of measurement), Joule had found for the power, P, of an electric motor:

$$P = iV \quad \text{and since by Ohm's law} \quad V = iR$$

$$P = i^2 R$$

Both of these equations are highly significant. In the first equation, power is seen as dependent upon two factors: (1) the current, expressing the movement of something, and (2) the voltage, expressing the gradient across which it is moving. The total effect, the total activity, is the product of the two. Power is energy divided by time, so that Joule had found that electrical energy (when energy it was called) can be factored into, or is dependent upon, a quantity of electrical charge moving across a difference of potential (voltage). Mechanical work had already been expressed in analogous terms, as a quantity of weight moved across a difference of height.

The second equation expresses the fact that electrical power is proportional to the square of the current. This is not intuitive since the current produced by the batteries is linearly proportional to the amount per unit time of their chemical reaction, as Faraday had noted. How can the rate of chemical reaction be proportional to the current, while the power is proportional to the current squared? This sounds like an amplification of power, a source perhaps of perpetual motion. The mystery is removed when it is recognized that the energy of the chemical reaction is the product of the amount of reaction times the chemical potential of the reaction—another case of energy being the product of a flow of something across a potential or gradient, just as with electrical energy or gravitational energy.

The second equation was significant in another respect, for Joule found that it applied to the electrical production of heat as well as the generation of mechanical power—that is, heat produced by the battery was proportional to iR^2. If mechanical power and rate of production of heat were both proportional to iR^2, then they must be proportional to one another: there must be an equivalence value relating a quantity of heat to a quantity of mechanical work.

Joule attempted to measure the heat equivalence of mechanical work by four different methods. In each case, mechanical work was supplied by the descent of weights and was measured in foot-pounds. Heat was produced by (1) electrical flow from a magneto generator, (2) viscous friction of water pushed through small capillary tubes by a plunger, (3) compression of a gas by a plunger, or (4) agitation of water by a paddlewheel (the method most commonly referred to today). Heat was measured by the rise in temperature of water using very precise thermometers that were employed in the brewery; Joule claimed he could measure to .005°F. The heat unit used by Joule was the British thermal unit (BTU): the amount of heat required to raise 1 pound of water by 1° Fahrenheit. Using the magneto generator, Joule got results as low as 587 or as high as 1040 ft-lbs per BTU, but, with the other methods, he gradually came to regard the correct value to be about 772 ft-lbs, which is close to the modern value of 778 ft-lbs. Since a BTU equals 252 calories of heat, and a ft-lb equals 1.356 joules of mechanical energy, Joule had found, in metric terms, that 252 calories of heat is equivalent to (772)(1.356) joules, or 1 calorie = 4.15 joules, close to the modern value of 4.18 joules.

It is worth noting that Joule determined, ultimately with great accuracy, the amount of heat that is produced from the dissipation of a given amount of mechanical work. He did not determine how much mechanical work is produced from heat, though he sometimes talked as if he had, supposing that the chemical action of his batteries was the same as heat. The amount of mechanical work that can be produced from heat is a more complex subject, which will be discussed in Chapers 4 and 5.

Joule presented his first results in 1843, but few paid any attention to them. He was an unknown amateur, the slight temperature differences he claimed to measure probably seemed unreliable to many listeners, and it was not "cool" to talk about heat as being manufactured by mechanical action when it was listed as one of the elements in Lavoisier's *Elements of Chemistry*. Not until 1847 did he find himself being heard, at a British Association meeting in Oxford, by many of Britain's greatest minds, including Michael Faraday and the young William Thomson (later Lord Kelvin). Thomson was skeptical but interested, and he later converted to Joule's view that heat and work are not individually conserved, but something related to their sum is—a something that Thomson called "energy." For Joule, mechanical work could not be destroyed without an equivalent agent of activity taking its place, for God would not allow it—for as he said, "the grand agents of nature are, by the Creator's fiat, indestructible."

HERMANN HELMHOLTZ

Like Robert Mayer, Hermann Helmholtz (or von Helmholtz as he was later titled by the Kaiser) thought deeply about philosophy as well as physiology and physics, and he was profoundly influenced by Kant's ideas about the interrelated forces of nature and about the sources of human knowledge in the mind's ordering of sensory experience. During his medical education, Helmholtz studied physiology with Johannes Müller (1801–58), who espoused a mechanistic view of life, but who was especially interested in sensory physiology, with all its deeply perplexing challenges to the meaning of experience and knowledge. Müller formulated a "principle of specific nerve energies" which states that the perceptions and responses elicited by a sensory nerve depend much more upon the central connections of the nerve than upon the mode of stimulation of its peripheral ends. The optic nerve, for example, can be stimulated by mechanical pressure rather than light, but we still see flashes of light because the signals are carried to the occipital lobes of our brain. If it were possible to graft the optic nerves to the auditory region of the brain, and the auditory nerves to the optic region, then we might hear the flash of lightning before we saw the thunder. As Wordsworth wrote in *Tintern Abbey*: "of all the mighty world of eye and ear— both what they half create, and what perceive." Müller's views of sensory physiology were in tune with Kant's ideas on the role of the mind in ordering sensory information. Helmholtz himself made great contributions to the study of both vision and hearing. Like John Dalton (who was color blind) and James Clerk Maxwell, he studied color vision, concluding that our perception of the entire spectrum of visual light results from only three types of color receptors in the retina. He also invented the ophthalmoscope.

It may seem odd that both Mayer and Helmholtz began as physiologists rather than physicists and that both were self-taught in physics. Partly it was a matter of money: a doctor could make a living, and Helmholtz received a free medical education by agreeing to serve as an army doctor for 5 years. In their medical studies, physiology gave them an entrance into the world of natural forces and change. But there was probably more to it than that. The living animal had long been at once an enigma and a refuge for explaining change. For 2,000 years, the living organism was a metaphor for all of nature: nature was alive, and all motion and change were testimony to that livingness. But gradually the role of machines in human life increased, and the metaphor began to change: nature became a machine. A living organism has a personality, a freedom, and is unpredictable, but a machine is determined, predictable, and follows rules of geometry and physics. In the time of Mayer and Helmholtz, the metaphor of the machine was being extended into the realm of life itself, and both Mayer and Helmholtz became proponents of a mechanical view of life. They believed that life operates within the same physical and chemical rules as the rest of nature, but, at the same time, it clearly shows a remarkable integration of mechanical, chemical, electrical, and thermal phenomena. If all these various forces are interrelated in life, and the total activity of animal life is measurable in terms of oxygen used, then perhaps the forces of all natural activity are interconvertible and measurable by a common standard. It is often supposed that advances are made first in the physical sciences and then applied to the study of living organisms. But this view is much too one-sided. The history of science is filled with examples of how the study of life has led to the expansion of physics and chemistry.

In 1847, when he was but 26 years old, Helmholtz published a long essay entitled *Ueber die Erhaltung der Kraft*, which translates to "on the conservation of force." A force (*Kraft* in German) meant a cause, and, for Helmholtz as for Mayer, and for Kant and Leibniz before them, a cause must equal its effect. Why? Because human understanding, the order of the mind, life itself, depends upon it, upon consistency in the way things work. For Joule, consistency and conservation of force (energy) and matter were part of God's creation, as they were also (more quietly) for Mayer. Helmholtz could invoke only rationality.

Force (*Kraft*) in Helmholtz's essay has two meanings, but it is clear by the context whether he is referring to a Newtonian force that acts centrally between two bodies or to what soon would be called energy. It is the *Kraft* of energy that Helmholtz argues is conserved.

Helmholtz presented his case for the conservation of energy by discussing many different phenomena, but he began with pure mechanics and the conservation of the sum of *vis viva* (kinetic energy) and what he called "tensions" (potential energy). As in free fall, for all cases involving central forces between bodies, if kinetic energy is gained, an equivalent amount of potential energy is lost, and vice versa:

> In all cases of motion of material points under the influence of their attractive or repulsive forces, of which the intensity depends only on distance, the decrease in tension always equals the gain in *vis viva*; and contrariwise, an increase in the former equals a loss in the latter. In other words, the sum of *vis viva* and tension is always constant. In this its most general form, we may designate our proposition as the law of the conservation of force.[1]

Helmholtz also applied to pure mechanics ideas that Sadi Carnot (1791–1831), in analyzing the action of the steam engine, introduced in his *Reflections on the Motive Power of Fire* (1824). Imagine a group of bodies that can, by their own central forces (tension) on each other, increase the motion of their parts and, in that sense, do work in going from their original state, A, to another state, B. If then the bodies are to be restored to their original state in order to repeat the process, a quantity of work equal to that gained in going from A to B must be put back in returning from B to A. (All this assumes that there are no frictional losses in either path of the cycle.) That the quantities of work must be the same, no more and no less, is necessitated by the postulate of the

---

1. Hermann Helmholz as quoted by Charles Coulston Gillispie, *The Edge of Objectivity*. Princeton: Princeton University Press, 1960, p. 390.

impossibility of perpetual motion, for if pathways between A and B existed that involved different amounts of work, it would be possible to rig the system so that one gained more work from A to B than was required to restore the system back from B to A. With this argument, Helmholtz has introduced (following Carnot) the idea of a state that has certain properties; in this case, the property of potentially doing work. If work is the only energy allowed in this particular mechanical system (heat being ruled out), then work can be considered a state function, and the amount of work obtained or required in going from A to B is the difference in the potential work values (tensions for Helmholtz) of the two states. It is like elevation on a landscape. If my house is 100 feet above the river, and I go down to the river, then I have to expect to climb a net 100 feet up to get back to my house, no matter what path I take. It would seem nice if the world were otherwise, for then we might arrange always to be walking (or better yet bicycling) downhill—but I expect we might have difficulty arranging where to meet with our friends.

The conservation of energy, E, has become one of the most secure foundations of modern science. Historically, it followed closely on the heels of the conservation of mass, m, and ultimately merged with it in Einstein's $E = mc^2$ equation (where c is the velocity of light), an equation that has been amply confirmed in the enormously energetic realm of nuclear reactions. When Henri Becquerel discovered (1896) radiations coming from uranium that were not powered by sunlight or any ordinary internal chemistry, and that seemed not to diminish with time, there was reason to doubt the conservation of energy law, but few did so, for it had so firmly won the confidence of nearly everyone in the 50 years since the young Helmholtz, Mayer, and Joule put it forward.

Can the law of the conservation of energy be proved theoretically from first principles? Apparently not. We have to accept the law itself as a first principle and be very glad that we have it—though it continues to rankle the theoretically inclined that its foundations are rooted in experience as well as in deductive logic.

In ancient Greece, the earliest philosophers conceived of a unified natural world by supposing that all things are made of one common substance. Thales chose water as that common matrix of nature, probably influenced by the observation that all life depends upon water. Others chose air or some undefined, indefinite substance. Empedocles decided that one nature required four elements: earth, water, air, and fire, each a unique representation of the opposite qualities of cold and hot, dry and wet—a unified scheme that survived, with various modifications, until the time of Lavoisier in the late 18th century. It was Lavoisier's genius not only to discover oxygen (contemporaneous with Joseph Priestley in England and Carl Wilhelm Scheele in Sweden) and to begin to construct our modern list of chemical elements, but also to place all chemistry firmly on a quantitative basis and to conceive that nature in all its changeability is unified by the conservation of tangible, weighable matter. The modern principle of the conservation of matter descends from the Ionian Thales's quest to find a tangible, material constant behind the natural world.

But behind the tangible matter is something intangible that creates change; the matter changes its position, velocity, form, and qualities even as its quantity remains constant. For nature to be consistent, for it to operate under a unified set of rules, the forces or causes of change must give the same results for a given set of conditions. In this sense, a properly defined cause has a definite effect, or, as Mayer, following Kant and Leibniz, said, a cause equals its effect. In a sequence of changes, this equality is passed from one stage to another, and, in the end, the imponderable something that causes change is said to be unchanged. But for this to be meaningful in quantitative science, the imponderable has to be measurable, and the common measure, created first for mechanical work (as the product of force times distance) and then for heat (as calories) has come to be called *energy*.

The concept of energy and its conservation unites all natural philosophy or science, facilitating predictions and problem-solving and setting restrictions on how anything can behave. Problems that are difficult to analyze otherwise may yield more easily with the further insight that energy must be conserved. The Earth in revolving around the Sun has a slightly elliptical rather than circular orbit, which means that it is sometimes slightly closer to the Sun, at other times farther away. When it is closer, the Sun–Earth system has less gravitational potential energy and therefore must have more kinetic energy—that is, the Earth's velocity in its orbit is greater when closer to the Sun and slower when it is farther away. The exact relation between the position of the Earth and its velocity can be most easily understood by invoking the conservation of energy.

The concept of energy unites all of nature, the living world with the nonliving. Living organisms are foci of activity and change, but they cannot create energy nor destroy it. It took humanity immense intellectual effort to come to the realization of that truth, that life is empowered and restricted by the same laws of energy as the rest of nature.

Chapter 2

Measurement, Dimensions, and Energy

THE MEANING OF QUANTITATIVE MEASUREMENT

By measure we mean a quantitative comparison of one thing to another of like character. Some things, like water, can be measured quantitatively; other things, like pain or happiness, cannot. To measure something is to compare it with a reference of similar nature. If one full glass of water can be poured into a second without reaching the second's brim, then the second glass has the larger volume, and we can make an ordered arrangement of volumes of glasses by this technique. But to make the measurement of volume truly quantitative, we need to find a small glass that can serve as a standard reference volume and count the number of times the reference glass can be emptied into the glass to be measured (never mind the errors of filling or emptying). The reference volume itself can be divided into parts, as in a graduated burette in a chemical laboratory. The ultimate measurement is expressed as a ratio of an unknown volume to the standard volume.

Accurate quantitative measurements go back at least to the 27th century BC, for the base of the Great Pyramid at Gizeh is an almost perfect square: two of its sides are within an inch of its average base length of 755 feet 9 inches. Accurate measurements of length do not require a modern scientific standard. The meter is no better than any other unit, including the ancient cubit which, like the foot, was derived from human anatomy, being the distance from a man's elbow to the tip of his extended index finger. Of course, a dependable unit has to be carefully defined (whose elbow and index finger are we talking about?), reproducible from one day and place to another, and agreed upon by all who use it. In the ancient world, at least eight different "cubits" were used in various places and times, just as different "gallons" are used even today. But the builders of the Great Pyramid were clearly all using the same reference standard of length.

Length was probably the most ancient of quantitative measurements, with weight, using the principle of the lever (proved formally by Archimedes), coming a somewhat distant second. Through geometry, measurements of length multiplied together could become measurements of area, or, multiplied again, measurements of volume. Measurements of weight, which in today's scientific terms are measurements of gravitational force, are taken in everyday life to be measurements of amounts of matter, or of mass, and the necessary insistence in physics that weight = mass × gravitational acceleration becomes an endless source of embarrassment and frustration on science examinations. The accurate measurement of short intervals of time came much later than that of length, area, volume, or weight, for the division of a day (solar or sidereal) into smaller parts was not easily standardized. Even the finest equatorial sundial was of no use on a cloudy day, could not be moved indoors, and could not be expected to measure the duration of the swing of a pendulum or the rolling of a marble down an inclined plane. Mechanical clocks were first constructed in the 13th century, but they were large and immovable and not very accurate, until great improvements were made in the 17th and 18th centuries.

It was Galileo who brought time into the laboratory by inventing a way to measure short durations—without any mechanical clock, metronome, or radio signal. He filled a large reservoir with water, with a tube at the bottom, from which water flowed into a small receptacle when he removed his finger from the end of the tube. Galileo wanted to measure the time required for a ball to roll various distances down an inclined plane. When he released the ball, he removed his finger from the end of the tube; when the ball reached a given mark on the incline, he closed the tube. How did he measure time? He weighed it! He inferred that the amount of time was proportional to the weight of water in his receptacle. The ratios of weights of water were compared to the ratios of distances along the incline, and, from these results, Galileo concluded that the speed of the rolling ball increased uniformly with time and not with distance, as he originally thought it should (it increases with the square root of distance).

Did Galileo really measure time by weighing water? There is a warning here concerning the limitations of human knowledge. The method not only assumes that water flows uniformly, but that time does as well—that time flows equably, as Newton later put it. How do we know that time never speeds up or slows down? We do not know, but the assumption feels right, and, more importantly, it is consistent with measurements of a great many motions in nature—at least until we measure balls rolling down inclines at nearly the speed of light, when we may have to make adjustments to either distance or time. Galileo's standard unit of time was a weight of water, perhaps that weight collected during a roll along the entire length of his incline. Clearly, his measurement of time was relative, not absolute. But that is hardly a limitation as there is no absolute measurement of time—nor of anything else. All measurements depend on a more or less arbitrary standard, whether it be the marks on a platinum-iridium bar in the National Bureau of Standards or the oscillations of cesium atoms.

Galileo expressed his results geometrically and in terms of proportions. Algebraic equations were only slowly penetrating mathematics in Galileo's time and were not yet used to express physical relations. The modern student becomes so familiar at an early age with algebraic formulas that he may not realize that there is a difference between a mathematical and a scientific equation. The former deals only with numbers, while the latter deals with numbers, with the standard units of measurement that influence those numbers, and with what is called the *dimensions* of those units.

DIMENSIONS AND UNITS

The dimension of a unit relates to the kind of measurement a unit is representing: length, for example, is given the dimension L, mass is given M, and time is given T. The dimensions for mass, length, and time suffice for the description of mechanical motion. They are the most common choice for describing mechanical phenomena, but not the only choice: force, length, and time are sometimes used by engineers; and force, length, and mass, or force, mass, and time would be possible (though clumsy) as well. Newton's second law (force = mass × acceleration) involves four variables: force, mass, length, and time, so that if three are stipulated, the fourth can be determined—hence only three fundamental dimensions are needed to describe motion. When thermal, electrical, or magnetic phenomena are considered, further dimensions are required.

Compound dimensions can be formed from the primary dimensions by multiplying or dividing one primary by another.

Area is length × length, and hence has dimension L^2, and volume has dimension L^3. Velocity is distance (length) divided by time, with dimension $L \div T$, or LT^{-1}, and acceleration is velocity divided by time, with dimension LT^{-2}.

A medieval philosopher might raise his eyebrows: what does it mean to divide distance by time? Can we also divide oranges by lemons? In geometry and the mathematics of proportions, ratios are formed only of like quantities. But in the transition from the arithmetic of numbers to the algebra of variables, a new twist is introduced, one that we seldom pause to think about.

In arithmetic, multiplication is like a repeated addition: e.g., four times five gives the same result as adding four fives together, or five fours. But in algebra, addition and multiplication are treated differently: the coefficients of $4x$ and $5y$ must be kept separate in addition, but they can be combined in multiplication: thus $(4x + 5y)$ cannot be further simplified, whereas $(4x \times 5y)$ yields $20xy$. This algebraic distinction between addition and multiplication is what we have when dealing with measurable quantities of different dimensions. If the variables x and y reflect quantities of things that have a different quality in measurement—a different dimension—then they cannot be added together. Yet multiplication is permitted for it is understood in the physical context that multiplication involves the creation of a new quality in measurement, a new dimension, from the preexisting ones—the creation of a dimension "xy" from "x" and "y." And thus length and time cannot be added together, or subtracted from one another, but they can be multiplied or divided.

Suppose we are in a car with a broken speedometer: What do we do to measure its velocity? We can measure the distance by counting the mileposts as our car passes them. We count the hours that pass on our watch. We divide the number of miles by the number of hours to get our average speed in miles per hour.

Arithmetically, what we have done is to divide one pure number (produced by the ratio of our distance to a standard mile) by another pure number (produced by the ratio of our total time to a standard hour):

$$\frac{\dfrac{l_x}{l_s}}{\dfrac{t_x}{t_s}} = v$$

where l_x = measured distance, l_s = standard distance, t_x = measured time, t_s = standard time, and v = velocity. The v in the preceding equation is a pure number, so we have not done anything here to bother the medieval philosopher. We have not divided oranges by lemons; we have just divided one pure number by another.

However, in an algebraic equation expressing velocity as a function of length and time:

$$\frac{l}{t} = v$$

the symbols represent variable numbers with dimensions (and standard units for those dimensions) attached to them. The equation is two relations in one, an arithmetical relation and a dimensional one, and with full disclosure would be written as:

$$\frac{l(L)}{t(T)} = v(LT^{-1})$$

In summary, the physical equation for velocity represents two relations in one: that of the pure numbers produced by measurement ratios and that of the dimensions of the things being measured (length, time) or compounded from those measurements (velocity). While the numbers follow the usual rules of arithmetic, the dimensions carry the restriction that they can be multiplied or divided, but not added or subtracted. The dimensions warn us that the numbers are influenced by the choice of measuring units. If a dimension is positive, then increasing the size of the unit for that measurement will decrease the number representing it. For example, if we have a length of 12 feet, the number decreases to 4 expressed as yards. In velocity, we have a negative dimension for time as well as a positive dimension for length. Increasing the size of the unit for a negative dimension has the opposite effect as that for a positive dimension. If the time unit for velocity is increased from minutes to hours, the number expressing velocity will increase by 60.

In pure mathematics, we deal with pure numbers that can be added together in any way we wish or operated on in various ways without restriction. In mathematics applied to science, however, the numbers are usually associated with dimensions of measurement, which introduces a restriction:

> If we wish to add or subtract quantities so as to form a homogeneous composite, then those quantities must have exactly the same dimensions. We would not wish to do otherwise— we would not add dissimilar quantities. We would not add 60 miles per hour to 30 square yards and call it 90 of anything.

Hence, in a scientific equation, if we have a number of terms, each term must have the same dimensions as every other term, and one side of the equation must have the same dimensions as the other. If, for example, $x + y + z = K$, then not only must x, y, and z each have the same dimensions, but so must K, as the equation could also be written as $x + y + z - K = 0$. (Zero itself, odd though it may seem, must be assigned the same dimension as the other members of the equation.). On the other hand:

$$\frac{(x+y+z)}{K} = 1$$

is an equation where both sides of the equation are dimensionless.

A scientific equation, then, must have dimensional homogeneity. If it does not, it will not be a valid equation if units of measurement are changed since the numbers on one side of the equation might be affected more by the change in units than those on the other side.

Using mass (m), length (l), and time (t) as the primary dimensional measurements in mechanics, here are the dimensions of the most common quantities used in mechanics:

Velocity, v	LT^{-1}	l/t
Acceleration, a	LT^{-2}	v/t
Momentum	MLT^{-1}	$m\cdot v$
Force, f	MLT^{-2}	$m\cdot a$
Pressure, p	$ML^{-1}T^{-2}$	f/l^2
Energy	ML^2T^{-2}	$f\cdot l$ or $p\cdot l^3$ or $m\cdot v^2$
Torque	ML^2T^{-2}	$f\cdot l$

Note that energy and torque, which are very different "things," have nevertheless the same dimensions in this scheme. Their dimensions could be distinguished, however, if the directions of the measurements of length were taken into account, for both lengths measured in the case of energy are in the direction of the force, while one of the lengths measured in torque is perpendicular to the force—that is, energy could be represented dimensionally as $ML_x^2T^{-2}$ and torque as $ML_xL_yT^{-2}$. Note also that the dimensions of energy correspond to those of force × length, pressure × volume, and mass × velocity squared.

In scientific equations, most variables—those innocent letters in the equation—carry dimensions associated with them, but some do not. Examples of dimensionless variables are angles, specific gravity, and strain. Angles—whether expressed in degrees or radians—are measured as the ratio of two lengths: the ratio of a circular arc length to the length of the whole circle, or to the length of the radius of that circle. The first ratio, multiplied by 360, gives degrees; the second ratio gives radians. It is sometimes stated that angles measured in degrees carry a dimension with them, but they do not. Specific gravity is measured as the ratio of two densities: the density of the object to that of water. And linear strain of an object under stress is measured as the ratio of two lengths: the length of the object's change to the length it has without stress. Angles, specific gravity, and strain are variables, but they do not have dimension.

Many constants have dimensions. The velocity of light in empty space is constant (independent, for example, of wave length, or of the velocity of its source), but it has the dimensions of any other velocity, LT^{-1}. The coefficient, G, in Newton's law of gravity is a constant throughout the universe (we suppose), but it has the dimensions $M^{-1}L^3T^{-2}$, as can be derived from the law, where f is the force, m_1 and m_2 are the masses, and r is the distance between them:

$$f = G\frac{m_1m_2}{r^2} \quad \text{or} \quad G = \frac{fr^2}{m_1m_2}$$

Hence G has dimensions $(MLT^{-2} \cdot L^2) \div M^2 = M^{-1}L^3T^{-2}$.

Some constants have no dimension. Examples are integers or any other pure numbers, including transcendental numbers such as π, which is the ratio of one length (the circumference of a circle) to another length (the diameter of the same circle), or e, the base of the natural logarithms, which is defined as:

$$\lim_{n \to \infty}\left(1+\frac{1}{n}\right)^{n}$$

where n is a pure number and so also is the limit.

In examining a scientific equation, we know that the following expressions are dimensionless:

1. Any angle, as stated above.

2. Any exponent. For example, in e^{x}, 2^{y}, and 10^{z}, x, y, and z must be pure numbers without dimension. This is evident when e^{x} is represented as a power series: $e^{x}=1+x+\dfrac{x^{2}}{2!}+\dfrac{x^{3}}{3!}+\cdots$ Since the first term on the right side of the equation is dimensionless, so must be all the other terms on the right in order to have dimensional homogeneity. Hence x itself must be dimensionless, as also e^{x}. Many scientific equations are exponential, as for example the equation for the decay of activity of a radioactive isotope: $A=A_{0}e^{-\lambda t}$. Since A and A_{0} have the same dimensions, $e^{-\lambda t}$ must be without dimension, and since λt is an exponent, it, too, must be dimensionless. The time, t, has dimension T, so the decay constant, λ, must have dimension T^{-1}.

3. Trigonometric expressions. Trigonometric functions such as $\sin(x)$, $\cos(x)$, and $\tan(x)$ represent the ratios of two lengths within a right triangle, and hence, like the angle x

itself, can have no dimension. Products, quotients, sums, or differences of these trigonometric functions are necessarily without dimension.

4. Any logarithm. Since exponents are without dimension, so must be logarithms. Consider the identity: $x = e^{\ln(x)}$. Since $\ln(x)$ is the exponent to which e must be raised to give x, $\ln(x)$ must be without dimension, and since e is dimensionless, the number x, whose logarithm is being taken, must also be without dimension. However, when logarithms are used as an alternative to exponential equations, as when $A = A_0 e^{-\lambda t}$ becomes $\ln(A) = \ln(A_0) - \lambda t$, we create an equation that seems to defy this last statement. The nondimensionality of λt reminds us that $\ln(A)$ and $\ln(A_0)$ must be dimensionless even if the quantities A and A_0 are not. Since we have logarithms on each side of the equation, they can be combined to form:

$$\ln(A) - \ln(A_0) = -\lambda t \quad \text{or} \quad \ln\left(\frac{A}{A_0}\right) = -\lambda t$$

The quotient A/A_0 is dimensionless even when A and A_0 individually have dimension. If a logarithm appears only once in an equation, then its operand must be dimensionless; but if the operand itself is expressed as a ratio, then two logarithmic terms can be formed whose operands may have dimensions even though the logarithmic terms themselves do not.

ANALYSIS OF DIMENSIONS IN SCIENTIFIC EQUATIONS

Analysis of dimensions can provide useful insights, not only in determining whether an equation makes sense (has dimensional homogeneity), but in predicting possible relationships among variables that may be contributing to a phenomenon under study. Consider the problem of free fall studied by Galileo. We now have cameras and electronic timers for doing the experiments, but before we spend a lot of time in the laboratory, a consideration of the dimensions of the possible factors involved may help us to predict the results. Let us try to write an equation that might show how distance fallen will vary with time.

In pure mathematics, we could write any relationship we wanted between two variables such as s and t. Maybe $s = t$, or $s = \frac{1}{2}t^2$, or $s = \ln(t)$, or $s = e^t$. But in the real world, the variables s and t stand for measured quantities that have dimensions attached to them: L for the distance and T for the time. There is then no valid equation that we can write between s and t alone for no equation using only those two variables will have dimensional homogeneity. Another factor (variable or constant) is needed to form an equation that will make sense—both now with whatever measuring units we are using and in the future if we change those measuring units.

What might a third factor be that will unite the dimensions of distance and time? To bring the dimensions of L and T together, the third factor must be a compound quantity that contains both L and T. But suppose we do not think of that right away and choose weight as a further variable that influences free fall. In Galileo's time, it was a common belief that heavier objects fall faster than light ones (they do slightly if air resistance is considered), so weight is a reasonable guess. But what are the dimensions of weight? Since weight is a force, it has the dimension of force, F, if that is considered a primary dimension, or (by Newton's second law) of mass × acceleration, MLT^{-2} if we use the mass-length-time system of units. If we regard force as a primary dimension, then we have only made things worse by adding weight to the pool of factors for we now have three variables each with its own dimension, with no possibility of making a dimensionally homogeneous equation from them. If we choose to regard mass rather than force as a primary dimension, then force is MLT^{-2}, which unites the dimensions of distance and time but, in doing so, introduces the new dimension M, which neither of the other variables contains. Again, we cannot write a dimensionally homogeneous equation using distance, time, and weight. What we need is a third factor that unites the dimensions of the other two without introducing any new primary dimension.

The most obvious variable that combines the dimensions of length and time without introducing a new primary dimension is velocity, v, with dimension LT^{-1}. Introducing velocity, we can write dimensionally homogeneous equations involving s and t, such as: $s = v \cdot t$ or $t = s \div v$, or $v = s \div t$. Each of these expressions is equivalent to the others, so we have obtained only one functional relationship among the variables, and that one is not very edifying, for we could have written it as a definition of velocity that applies to any motion and hence says nothing of specific interest about free fall.

But if we go the next step and choose acceleration as the factor to bring distance and time together, we get something more interesting. Acceleration has dimensions LT^{-2}. We can find out how to relate it to time and distance by the following systematic method. We have three variables, s, t, and a, and two primary dimensions, L and T. If we let s be proportional to t^x and to a^y, where x and y are powers (exponents) of their respective factors, then we can write the following equation:

$$s = k \cdot t^x \cdot a^y,$$

where k is a dimensionless number, a coefficient of proportionality. (For example, the area of a triangle is proportional to both its base and its height, with a proportionality coefficient of ½.) We have two unknowns, x and y, in our equation, but the two primary dimensions, L and T, allow us to write two further equations, based on the requirement of dimensional homogeneity, to find the two unknowns. Count up the powers of L and T on each side of the preceding equation. The coefficient k has no dimension, so we can ignore it. Dimensionally, the factors in the equation are:

$$(L) = (T)^x + \left(LT^{-2}\right)^y$$

$$\text{For } L: \quad 1 = 0 + y$$

$$\text{For } T: \quad 0 = x - 2y$$

From the first of the dimensional equations, $y = 1$, and, from the second, $x = 2y = 2$. With the exponents x and y now determined, the preceding equation becomes:

$$s = kt^2 a$$

which tells us that the distance fallen is proportional to the square of the time, which is what Galileo found experimentally. If acceleration is constant, as it is pretty nearly on the surface of the Earth, the coefficient k, which we cannot find by dimensional methods alone, is ½, and the equation becomes $s = \frac{1}{2}at^2$. This familiar equation is easily (more easily you may say) derived by other methods, but the dimensional method provides a check (except for the coefficient ½) and an aid to memory and understanding. It would have been an aid even to the great Galileo!

When the problems get a little more complicated than the simple one of free fall, an analysis of dimensions can become a great help. Consider the swing of a pendulum. What determines the period, p, of its swing? We might guess that the pendulum's mass, m, its length, l, and the gravitational acceleration, g, are the determining factors—but in what way do they determine the period? Let the equation be:

$$p = k \cdot m^x \cdot l^y \cdot g^z$$

which becomes in terms of dimensions (ignoring k)

$$T = (M)^x + (L)^y + \left(LT^{-2}\right)^z$$

Since there are three primary dimensions, we have three equations:

$$L: \quad 0 = 0 + y + z$$
$$M: \quad 0 = x + 0 + 0$$
$$T: \quad 1 = 0 + 0 + 2z$$

In the second equation, we see that $x = 0$, which says that mass has exponent 0, so that the mass of a pendulum does not affect the period of its swing, contrary to our initial guess. From the third equation, $z = -\frac{1}{2}$, and from the first equation, $y = -z = \frac{1}{2}$. Our derived equation for the swing of a pendulum is then:

$$p = k \cdot \left(\frac{l}{g}\right)^{\frac{1}{2}}$$

Intuition alone would not have told us that the period of a pendulum is proportional to the square root of the pendulum length, inversely proportional to the square root of gravitational acceleration, and independent of mass (or weight) of the pendulum, but dimensional analysis easily yields these results. Dimensional analysis does not tell us whether or not the period is influenced by the amplitude (angle) of the swing because the angle is necessarily dimensionless. Experimental measurements, or a more detailed physical analysis, shows that the period is nearly independent of the angle for small angles but not for large.

As a final example of the utility of thinking about dimensions, consider the Pythagorean theorem. Area has dimension of L^2. Geometric figures that are similar in shape have areas that are proportional to the squares of their corresponding linear sides. This is obvious for squares of different sizes and soon becomes clear for triangles and other figures. In a right triangle, a line drawn from the apex of the right angle perpendicular to the opposite side divides the original triangle into two triangles that are similar to each other and to the original triangle (parent and daughter triangles all have the same three angles). The area of each triangle is proportional to the square of its hypotenuse. But the hypotenuse of a daughter triangle is a side of the parent triangle, the sum of the areas of the daughter triangles is the area of the parent, and the proportionality constant relating area to the square of the respective side is the same for all similar triangles. Therefore the square of the hypotenuse of the original triangle equals the sum of the squares of the other two sides. There are scores of other proofs of the Pythagorean theorem, but this one, involving dimensional thinking, is perhaps the simplest.

THE DIMENSIONS OF ENERGY

Dimensions are also very helpful in understanding the various forms of energy. In purely mechanical systems, which are measured using the primary dimensions of M, L, and T (or F, L, and T), the dimensions of energy, ML^2T^{-2}, can arise in different contexts:

1. As a force acting through a distance: $(F)(L) = (MLT^{-2})(L) = ML^2T^{-2}$

2. As a surface tension (a force per length) acting through an area: $(FL^{-1})(L^2) = (MT^{-2})(L^2) = ML^2T^{-2}$

3. As a pressure (a force per area) acting through a volume: $(FL^{-2})(L^3) = (ML^{-1}T^{-2})(L^3) = ML^2T^{-2}$

4. As a force acting through time to accelerate a mass through a distance (such as happens in free fall). The mass acquires increasing velocity, so that its momentum (mass × velocity) increases. As its velocity increases, so also of course does its velocity squared, and mass × velocity^2 is proportional to energy, as seen in its dimensions: $(M)(LT^{-1})^2 = ML^2T^{-2}$. Uniform acceleration, as in free fall, can be viewed (though it may seem a strange view) as increments of impact (force × time)—or increments of momentum resulting from the impacts (note that impact has the same dimensions as momentum: MLT^{-1})—acting on velocity, so that we have $(MLT^{-1})(LT^{-1}) = ML^2T^{-2}$. This form of energy was called *"vis viva"* (living force) by Leibniz, but renamed *kinetic energy* by William Thomson (Lord Kelvin). We will discuss later why kinetic energy goes by the familiar form of $\frac{1}{2}mv^2$ rather than just mv^2.

When energy is measured outside the realm of pure mechanics (as in thermal, chemical, electrical, and magnetic phenomena) further primary dimensions are required (such as temperature, electrical charge, and magnetic field). These lead to additional ways of expressing the measurement of energy.

Chapter 3

The Laws of Energy

CHANGE INVOLVES CURRENTS MOVING ACROSS GRADIENTS

When we inquire what causes anything to change, we can start with this very simple rule: whenever in the world of nature we see something happening, there also we see the existence of differences. In truth, we may not immediately "see" the differences, but they can be found if we look hard enough for them. If everything were the same, nothing would happen. If all water is at the same level, we do not see water flowing—we have a placid lake rather than a running stream. It takes a difference to make a waterfall: a difference of height. If Lake Ontario were suddenly to be at the same level as Lake Erie, there would be no Niagara Falls, and all the tremendous power we associate with Niagara—all the electrical conduits stretching out across the northeastern countryside—depends on that one difference: 167 feet of height and water flowing over it. I was riding the electric trolley in Boston in 1965 when the electrical flow from Niagara broke down. All the lighting of houses and streets in Boston and New York went out, along with the electric gasoline pumps at service stations, the elevators of office buildings, and much of the medical equipment in hospitals. All this and more was driven by a flow of water sustained by one difference in height and a subsequent flow of electricity sustained by a difference in voltage generated by that flow of water.

All natural change can be broadly viewed as a flow, a flux, a current, of "something" moving across a difference, or, as we will call it, a gradient. Where electricity flows through wires, there must be a gradient of electrical potential (voltage). Where gases escape from jets, or wherever indeed the wind blows, there must be a gradient of pressure. Where a hot body warms a cold one, the flow of entropy (commonly called "heat") is sustained by the temperature gradient between the two bodies. Wherever a chemical reaction occurs (and this includes the vast array of chemical reactions in the living organism), there must be gradients of chemical potential created by differences in chemical composition and/or in chemical concentrations. Not a single exception to this rule can be found, and this is as true for living things as for the rest of nature: all change depends on the existence of gradients of various kinds, and all change involves flows across those gradients. The fluxes may be macro in scale, where we can directly see them, or submicro, where we cannot, but there is always motion, whether it is in the nucleus of an atom, the atom itself, the molecule, or the river.

Gradients and their respective flows come in various forms, but one type of flow can often be coupled to another, and, from this coupling, one form of gradient may generate another. A flow of water, for example, can be made to create an electrical gradient and electrical flow (with a generator), or an electrical flow can create a pressure gradient and water flow (with an electric motor and pump). Electrical flows can create chemical flows (reactions), as in the extraction of aluminum from its ore or in the charging of a battery, or chemical flows can create electrical flows, as in the discharging of a battery. The discovery that one form of activity or change can be converted to another gave rise to the idea of a commonality of the different forms—that is, to the concept that a common energy lay behind all the various forms of natural activity.

THE INTENSIVE AND EXTENSIVE FACTORS OF ENERGY

How does one measure an amount of activity or change? As Joule found with his electric motor, its power was proportional to both the current and the voltage across which the current was flowing from his batteries. So it is with all currents of activity: we need to recognize both the current and its gradient and multiply the one by the other. Compare Yosemite Falls with Niagara: Yosemite's gradient is much greater than Niagara's, but its current is much less. It is because the whole of Lake Erie flows out (in 3 years) across Niagara that its activity is so much greater than that of a small stream falling from many times its height. The common measure of the various forms of activity (as we saw in Chapter 1) is energy, and energy can be factored into two terms: an intensive (potential) term, such as the height of the waterfalls, and an extensive (capacity) term, such as the water falling across the falls.

Table 3.1 lists the intensive and extensive factors for several forms of energy. An extensive factor depends on the extent, or scale, of an object or system considered, while an intensive factor remains the same with changes in scale. For example, if a liter of air is split into two half-liters, volume is diminished but pressure remains the same.

Table 3.1 THE MEASUREMENT OF ACTIVITY OR ENERGY

Type of Activity	Intensity Factor	Capacity Factor	Energy Unit Conversion
Gravitational	Height (meters)	Weight (newtons)	meter-newton = 1 joule
Electrical	Voltage (volts)	Charge (coulombs)	volt-coulomb = 1 joule
Pressure-Volume	Pressure (atmospheres)	Volume (liters)	atmos-liter = 101 joules
Thermal	Temperature (degrees)	Entropy (calories/degrees)	calorie = 4.18 joules
Chemical	Chem. Potential (calories/mole)	Mass (moles)	calorie = 4.18 joules
Diffusional	Diff. Potential (calories/mole)	Mass (moles)	calorie = 4.18 joules
Kinetic	Velocity (meters/sec)	Momentum (newtons-sec)	meter-newton = 1 joule

Energy can be calculated, whatever its form, as the product of two factors, one "intensive" (the potential or difference or gradient), the other "extensive." Thus height, voltage, pressure, temperature, and chemical potential are intensive (or intensity) factors, and weight, charge, volume, entropy, and mass (or moles) are extensive (or capacity) factors. Intensive properties are ones that are not changed by dividing a body with that property, while extensive properties are diminished by such division. Thus, for example, 1 liter of gas at atmospheric pressure could be split into 2 half-liters still at atmospheric pressure; the capacity factor (volume) is changed by such division, but not the intensity factor.

The gravitational capacity factor is listed here as weight. Weight is a force: the force of gravity on the object. Since force seems more like an intensity factor or potential than a capacity factor, some authors reverse the terms and list weight as the intensity factor and height as the capacity factor—but this, too, creates a strange perspective. The difficulty can be overcome, for the purist, by recognizing that weight is mass × gravitational acceleration. Take the gravitational acceleration away from the weight (by division) and put it with the height (by multiplication). That gives a "gravitational potential" (related to distance or height) as the intensity factor and mass as the capacity factor, thus making the factors for gravitational energy more truly equivalent to those for the other forms of energy.

THE FIRST LAW OF ENERGETICS:
THE CONSERVATION OF ENERGY

The last column of Table 3.1 expresses the fact that when one form of energy is completely transformed into another, a given quantity of the first produces a definite quantity of the second—no more, no less, and always the same. It is like a currency exchange of dollars for pounds that is not only regulated, but absolutely so for all time and without any commissions allowed. We never get too much or too little, which leads to the conclusion that the totality of energy is conserved. This is the meaning of the first law of energetics (or thermodynamics): that in the transformations of one type of activity into another, total energy is conserved. Total energy cannot be created, nor can it be destroyed.

The conservation of energy is of profound significance for all areas of pure and applied science. No longer need anyone wonder whether a machine can be built that will create its own energy, no matter how complicated and incomprehensible the machine may be to the outside viewer. A friend (after a few too many drinks) remarked that our world needs to figure out how to amplify energy. But that, the first law says, is impossible. Save your time and investment. Figure out how to utilize more of the energy of the Sun, but do not try to make energy multiply itself. Even Einstein, who was very good at being skeptical about the foundations of physics, thought that the laws of energy are as certain as anything in science.

In physics, one can be assured that no matter how surprising, unfamiliar, or complicated a phenomenon may be, the conservation of energy holds true. Becquerel's discovery (1896) of uranium's radioactivity brought about the recognition of a new form of energy (nuclear), but it did not raise doubts for long that total energy (or, in this case, energy plus mass) is conserved. The analysis of many mechanical systems, such as the swing of a pendulum or the vibration of a spring, is facilitated by the certainty that the combined kinetic and potential energy of the system remains constant if friction is absent.

In chemistry, the conservation of energy is of huge importance, as we will see in Chapter 7. It is the basis for the construction of tables of state functions (internal energy, enthalpy, entropy, free energy) that permit one to predict whether or not a chemical reaction can proceed under prescribed conditions.

We have seen that the conception of the conservation of energy by Mayer and Helmholtz was related to their conviction that living organisms obey the same principles as the rest of nature—and that the union of forces that they could see in life must also happen throughout the entire natural world. For the future of biology, this union of life with physical and chemical principles was as revolutionary as the simultaneous development of Darwin's theory of evolution by means of natural selection. No longer could life be viewed in animistic terms as a source of energy. Living things cannot create their own energy, but rather they are dependent on the energy sources of the world around them. The activity of life is not self-sustained, but depends on outside gradients. That may be humbling. It may also be beautiful. As the poet John Donne wrote, "No man is an island entire in itself, each is a part of the main, a part of the continent."

THE INCREASE AND DECREASE OF
POTENTIALS AND THEIR ASSOCIATED ENERGY

We have noted that behind every change is a gradient. We can say that the gradient is the cause of the change, so long as we recognize that it is a necessary but not a sufficient cause, that the gradient may exist for a long time before anything happens—as a test tube of hydrogen gas can exist in a room full of oxygen until someone lights a match.

But if the cause of every change is a gradient, the effect of every change is the reduction of the gradient giving rise to the change. As a stream flows down a mountainside, it erodes the mountain and decreases the gradient that causes the flow. As a gas flows from a high pressure to low, it tends to equalize the pressures. As electrons flow from high density (voltage) to low, they destroy the voltage difference that caused them to flow. There is no exception throughout nature to this rule: where a change is occurring, a gradient is being reduced.

The intensive factor, or potential, of any energy reservoir is increased or decreased by the addition or removal of the substance-like extensive factor (which we will often call the "charge," generalizing from electricity)—the potential is enhanced by charge flowing into it and diminished by charge flowing out. The pressure in a balloon or tire is increased as a volume of air is blown or pumped into it, and the pressure goes down if the balloon or tire lets air leak out. The voltage of an electrical capacitor is increased as charge is added to it, and the voltage diminishes as the capacitor discharges. A chemical potential is increased as the concentration of reactants is increased and decreased as the reactants are used up. A rechargeable battery, which is an electrochemical system, loses its voltage as its chemical reactants are used up in discharging, and it regains its voltage when the reactants are restored in recharging. Temperature goes up (usually) when entropy flows into a body and goes down (usually) when entropy flows out (exceptions occur in changes of phase).

The capacitance of an energy reservoir is the amount of charge it must receive (or lose) for a given increase (or decrease) in its potential. For example, the volume capacitance, C_V, of a tire is its change in volume, ΔV, divided by its change in pressure, ΔP: $C_V = \dfrac{\Delta V}{\Delta P}$. Likewise, the charge capacitance, C_Q of an electrical capacitor tells how much charge, ΔQ, is required to change the voltage by an amount ΔV_e: $C_Q = \dfrac{\Delta Q}{\Delta V_e}$. In a graph of charge against potential, the capacitance is the slope, and *if* the capacitance is constant, the graph is a straight line. The area under this line between any two potentials is the energy that has been added with the addition of charge, and this area is a triangle if capacitance is constant. For simplicity, consider the case of starting at zero potential (with zero charge) and filling the energy reservoir with the addition of charge to a final state of charge Q and potential V. Since the area of a right triangle is one-half the product of its two sides, the graph shows that the total energy, E, of the system is then:

$$E = \frac{1}{2}QV = \frac{1}{2}CV^2 \ \text{ (because } C = \frac{Q}{V} \text{ or } Q = CV\text{)}$$

Note that if energy is expressed as a function of potential alone, it is proportional to the square of the potential, and note also the "integration factor" of one-half. At each step in the addition of charge, the added charge represents a greater increment of energy as the charge is added against a greater potential. The total energy at the end of charging is not the product of the final charge and final potential, but exactly one-half that value—provided that the capacitance is constant (in thermal energy, it is not constant for more than a limited range of temperatures).

The equation relating energy to capacitance and potential looks a great deal like the equation for kinetic energy, $E = \frac{1}{2}mv^2$. The two cases are in fact analogous for the capacitance in kinetic energy is the momentum, and the potential is the velocity. This may sound strange. The whole idea of treating kinetic energy like the other cases may seem odd. How do we factor mv^2 into extensive and intensive factors and treat it as if the potential of kinetic energy is built up by a flux of the extensive factor? There are only two possibilities for factoring mv^2: either into $(m)(v^2)$ or into $(mv)(v)$. In either case, the factor containing m must be regarded as the extensive factor and that containing v (or v^2) the intensive factor. But only in the second case will addition of the extensive factor increase the intensive factor for increase in mass alone cannot increase the velocity of a body, but increase in momentum will. Thus the extensive factor of mv^2 is mv, and the intensive factor is v. The action of a sustained force upon a body over time gradually increases its momentum, and, in doing so, since the mass remains constant (at modest speeds), drives the body to higher velocity and higher kinetic energy. The capacitance of the kinetic energy system is its extensive factor divided by the intensive factor, which is momentum divided by velocity, which is the mass—mass is the amount of momentum that a body can hold at a given velocity. Since the capacitance (the mass) remains constant, the final kinetic energy is one-half the extensive factor (mv) times the intensive factor (v), or $\frac{1}{2}mv^2$, exactly similar to the charging of an electrical capacitor or of other energy reservoirs.

In summary, energy sources and their potentials diminish as their action proceeds. In this discharge of charge and energy, the following are important to note:

1. Charge moves from higher potential, P_1, to lower potential, P_2, and tends to reduce the potential difference and energy in doing so.

2. For a given quantity of charge, Q, moving across the potential, the energy, E, involved is: $E = Q(P_1 - P_2)$. For every

unit drop in potential, the energy that Q carries with it is the same, and thus if Q moves across two units of potential instead of one, twice the energy is involved. It is like a pound of water falling across 2 feet instead of 1: it can do 2 foot-pounds of work instead of 1.

3. From the graph of charge against potential, in which the area under the line is energy, it is evident that a greater increment of energy is required to raise the potential by one unit at higher potential than at lower potential. This is also evident from the equation that shows that, for constant capacitance, the energy is proportional to the square of the potential. This simple fact had consequences in the history of the analysis of the steam engine, where it was noted that it took more fuel to raise the temperature by a degree at high temperature than at low.

4. If the highest potential of an energy system is P_1 and its lowest potential is zero, then a quantity of charge, Q, moving across the entire potential carries $(Q)(P_1)$ of energy with it. If the same charge can move only across a potential difference of $(P_1 - P_2)$, then it carries only $(Q)(P_1 - P_2)$ of energy. The ratio of the energies for these two cases is: $\dfrac{Q(P_1 - P_2)}{Q(P_1 - 0)} = \dfrac{(P_1 - P_2)}{P_1} = 1 - \dfrac{P_2}{P_1}$.

If this ratio is regarded as an efficiency of utilization of the total energy that might be had from the movement of charge Q, it is evident that this efficiency increases as the value of P_2 decreases—that is, as the potential difference utilized is a greater part of the entire potential of the system. If a rock is let down by a rope and pulley from a height of 10 feet, but is allowed to fall only 2 feet, it can do only 20% of the work which it could do if it were allowed to fall the entire 10 feet.

5. In the discharge of an energy system, the same amount of energy can be used in moving a small charge across a large potential as in moving a large charge across a small potential: if $(Q_1)(\Delta P_1) = (Q_2)(\Delta P_2)$, then the energies are the same. In coupling two energy gradients together, the first in discharge while the second is charging, it is possible to create a higher potential in the second than is present in the first, but with a smaller flow, or a higher flow in the second than is present in the first, but with a smaller potential. This is the case, for example, in step-up or step-down electrical transformers, in which 110 volts can be transformed to 220 volts, or vice versa.

In many energy systems, the entire potential of the system can be utilized. For example, the charge from a battery can move across the entire voltage of the battery to a voltage of ground zero, or a volume of gas can move from its original pressure to a near zero pressure. But, in some cases, the earthly environment in which we work prohibits reaching a zero potential. This is especially true for thermal energy, which is limited by our ambient temperature being far above the absolute zero of temperature. The efficiency of a thermal system is limited by the temperature gradient that can actually be achieved compared to that which would be available if the Earth were (heaven forbid!) at absolute zero temperature. Gravitational energy is another case, but we seldom worry about it. We define whatever earthly surface we want as zero and calculate the height potential from that. Because the distances measured are usually small compared to the nearly 4,000 miles to the Earth's center, the gravitational force does not change significantly, and the energy available is the product of the height times the weight without any integrating factor of one-half.

FREE ENERGY AND BOUND ENERGY: THE SECOND LAW OF ENERGETICS

If energy is conserved, as the first law says, then why cannot a living organism or a nonliving machine, once it gets hold of some energy, use the same energy over and over? Why must it always be dependent—as it is—on its environment for new supplies of energy? The answer is that the living organism or the working machine requires free energy, which is energy that can do work, that can cause change, and that is associated with gradients and currents across those gradients, as discussed earlier. But the currents across the gradients gradually destroy the gradients, and a given source of free energy approaches an equilibrium in which there is no longer a gradient and hence no longer any free energy.

The discharge of free energy across a gradient may do either of two things. It may do work in building up a different gradient, in which case we say that the loss of free energy of the first gradient is compensated by the gain of free energy of the second. For example, the gradient provided by the oxidation of our foods can drive the synthesis of proteins, nucleic acids, and other molecules needed by our cells. Or, second, the free energy may not do work, may not be compensated by the building of another gradient, and may be lost as free energy. What has happened to this lost free energy? It has become converted to an energy that has no gradient but is uniformly dispersed as microscopic thermal activity. It is, to refer again to the oxidation of our food, the energy that keeps us warm, but since it is uniformly distributed, it can do no work. J. N. Brönsted called it equipotential energy, or energy that exists in a system that has no differences of potential. Helmholtz called it *bound energy*, a term that is no longer commonly used, but which is vivid and concise.

We experience the conversion of free energy to bound energy most commonly and directly in the warming effects of friction or impact, but the conversion also happens in the dispersion or mixing of materials or in the expansion of volume—processes that may not result in any increase in temperature.

The second law of energetics (thermodynamics) states that free energy can be converted to bound energy, but not the other way around. If an isolated system has a gradient within it, then it has free energy that can drive a process of doing work and/ or of creating bound energy. But if an isolated system has no gradient, its energy is entirely bound; the bound energy cannot by itself be converted to free energy or do any work. An isolated gradient can diminish its free energy but cannot by itself increase its free energy. If two blocks of metal, one hot and one cold, are brought together, the thermal gradient allows the flow of entropy from the hot block to the cold one, but not the other way around, and, once the two blocks have come to the same temperature, the thermal gradient will not, without outside intervention, be restored.

The second law was understood—or let us say, taken for granted—at some level long before the concept of energy and the first law were formulated for it is intimately related to one's expectations about the sequence of events in time. It is, as Arthur Eddington called it, "time's arrow"—it points the direction of time and of change. One could hardly expect a lake at the base of a mountain to start emptying itself by flowing up the ravines of the mountain above it. I looked down once from the top of Black Down in England to the pastoral farmland below and watched the smoke curling upward from the chimneys of the cottages.

I should have been more than astonished if the smoke instead had gathered itself from the sky and flowed downward into the chimneys, to combine with the ashes in the cottage fireplaces to form new logs set beside the hearths. A world would make no sense without the second law for the future would be indistinguishable from the past. The second law implies that nature is historical, that it moves in linear (not circular) time.

The second law is usually stated, as Rudolf Clausius defined it in 1865, in terms of the tendency of entropy to increase. We will look at entropy more carefully in the next chapter, but we do not need it to state the second law, which can be defined more simply as the tendency for gradients to be destroyed and for free energy to diminish toward zero. I wanted to say that the second law can be understood more simply by this means, but "understanding" itself is a slippery concept. Both the first and second laws of energetics are inducted from our experience and cannot be derived from pure theory or pure reason.

EXAMPLES OF THE CONVERSION OF FREE ENERGY TO BOUND

Suppose we have a car battery, a heating coil, and a room. The battery is used to heat the coil, which heats the surrounding room. Prior to using the battery for heating, the battery could have been made to do any of a number of things by connecting it to an electric motor, or it could have powered a radio set or any number of electronic devices (in the biochemistry department at Oxford University long ago I used a spectrophotometer powered by a car battery). Clearly a battery is a source of free energy. But after heating the room, the discharged battery and the warmer room can do nothing whatever because the warmth produced by the battery and electric coil has been evenly dispersed about the room, and there are no longer any gradients—chemical, electrical, or thermal. All the energy has become equipotential (Brönsted) or bound (Helmholtz). The chemical gradient inside the battery has been destroyed, and no new gradient has taken its place. Although no loss of total energy has occurred (chemical energy having been converted to thermal energy), the free energy of the battery has become the bound energy of the warmer room. And we cannot have it the other way around: no matter how long we wait, the warm room will never recharge the battery for us. Uniformity can never create gradients, nor bound energy convert itself to free.

As a second example, consider a cylindrical tank full of compressed air. The high pressure in the tank is due to the kinetic energy of the individual molecules of nitrogen and oxygen, which bombard the walls of the tank. Clearly there is energy (activity) in the tank. But how much of it is free? Considered as the tank alone, none of its energy is free, for there is no gradient within the tank, and no work can be done. But considered as the tank within a room, a gradient exists between the high pressure in the tank and the atmospheric pressure in the room, so that free energy is present, work could be done, and change could be manifest—if the domain we are considering is expanded from the tank alone to the tank plus the room. If the valve of the tank is opened, the air rushing forth from the tank could be made to do work, as by driving a little windmill or turbine. But as the tank empties, its pressure drops, and, ultimately, its pressure comes to that of the room itself, and no more work can be done as there is no longer a gradient. If the room is airtight, its pressure will have increased slightly; no energy will have been lost from our system, but all the free energy will have been converted to bound.

A QUANTITATIVE EXAMPLE OF FREE ENERGY

As we have just noted, a gas at high pressure can do nothing unless there is a region of lower pressure into which it can expand. If there is such a region, then there is a heterogeneity in our system, a difference of pressure, which can be a source of work (change) as gas flows from the region of higher pressure to that of lower pressure—the gas could drive a turbine, generate electricity, push a plunger to compress a spring or raise a weight, and so forth. Free energy is associated with the difference in pressure, but as gas flows from high pressure to low it diminishes the former and increases the latter, destroying the difference. The free energy will all be converted to bound energy unless the gas flow is coupled with another process (such as the lifting of a weight by the turbine) that replaces the original gradient with a new one.

Suppose that we have 1 mole of an ideal gas in a closed cylinder, but with a piston that can slide in or out at one end of the cylinder. If the pressure on the outside of the piston is equal to that in the gas, the gas cannot expand; but if the outside pressure is reduced a little, the gas will push the piston outward (assume no friction). Let the initial pressure of the gas at position 1 be P_1, its volume V_1, and its concentration C_1 ($C_1 = 1/V_1$ because there is 1 mole of gas). Let the gas expand to a greater volume by pushing the piston outward to a position 2, where it has pressure, volume, and concentration of P_2, V_2, and C_2. In pushing the piston, the gas might do work on it, converting its own free energy to a new free energy of the piston in some way. For example, suppose that the pressure on the outside of the piston is due to a weight that the piston moves upward as the gas pushes outward. Then free energy of the gas will be transferred to free energy of the weight as it moves upward.

Whether or not the gas does work on the piston as it expands, there are things happening to the gas: its volume gets bigger, its concentration gets smaller, and its pressure diminishes. (With the decrease in pressure, the resisting weight on the piston must be made smaller if the expansion is to continue.) If the gas does work on the piston, energy is transferred from the gas to the piston, and the energy of the gas decreases. Since the energy of an ideal gas is linearly proportional to its temperature, the loss of energy causes the gas to cool. But suppose that this cooling is countered by allowing thermal energy to flow across the walls of the cylinder into the gas from the environment, so that the temperature of the expanding gas remains constant (the gas expands "isothermally"). Then the energy of the gas remains constant with the temperature as thermal energy of the environment is exchanged, in effect, for work.

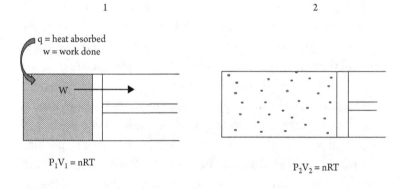

1   2

q = heat absorbed
w = work done

W

$P_1V_1 = nRT$

$P_2V_2 = nRT$

Although the total energy of an isothermally expanding gas remains constant during an expansion, the free energy of the gas decreases. The expanded gas can no longer do something that the original gas could do: push the piston outward from position 1 to position 2 against a resisting pressure or weight. The free energy lost is the maximum work the gas could do on the piston in moving it from 1 to 2. This maximum work is the product of the maximum force that it can exert times the distance that it moves the piston with that force:

$$\text{work} = (\text{force})(\text{distance})$$

The force that the gas can exert is its pressure times the surface on the piston:

$$\text{work} = (\text{pressure})(\text{surface})(\text{distance})$$

And the piston's surface times the distance that it moves is the volume displaced by the piston as it moves, that is, the volume of expansion of the gas:

$$\text{work} = (\text{pressure})(\text{volume of expansion})$$

Thus the maximum work done by the gas as it expands is the product of its pressure times its change in volume. But as the volume of the gas gets bigger, the pressure of the gas gets smaller. Our ideal gas obeys the ideal gas law, $PV = nRT$, where n is the number of moles (1 in this case), R is the gas constant, and T is the absolute temperature. Since n and R are constants, and T is too because the expansion is isothermal, PV has a constant value throughout the expansion and:

$$P = \frac{RT}{V}$$

For each little increment of expansion, the work that could be done by the gas is:

$$\text{work} = P\Delta V$$

To determine the entire work done in expanding from position 1 to position 2, we must add up many increments of $P\Delta V$, so that we can allow for the decreasing P as the expansion proceeds. Since P is, by the ideal gas law, a known function of V (as above), we can employ the calculus to do this summation (integration) of work increments over the entire expansion from V_1 to V_2:

$$work = \int_{V_1}^{V_2} PdV = RT\int_{V_1}^{V_2} \frac{1}{V}dV = RT\ln\left(\frac{V_2}{V_1}\right) = RT\ln\left(\frac{C_1}{C_2}\right)$$

That $\dfrac{V_2}{V_1} = \dfrac{C_1}{C_2}$ is evident from the fact that both $(V_1)(C_1)$ and $(V_2)(C_2)$ represent the constant number of moles (1) in the gas, so that they are equal to each other.

Since the maximum work that the gas can do in its expansion is work that the expanded gas can no longer do, it is exactly equal to the loss of free energy by the gas in its expansion. Therefore, the free energy of the gas has decreased by:

$$\text{free energy loss} = RT\ln\left(\frac{C_1}{C_2}\right) = RT\ln C_1 - RT\ln C_2 \qquad (3.1)$$

Equation 3.1 is of enormous significance in chemical and biological energetics. Although it was derived from the maximum work that a gas can do in expanding, it applies even when no work is done. The gas in the greater volume of position 2 does not know whether it has done any work in expanding from position 1. It is described by its present state through the ideal gas law, which can be called its "equation of state." If the gases of 1 and 2 are at the same temperature and have the same number of moles (1 in this case), then they have the same total energy, expressed as pressure-volume energy, $(P)(V)$. But the more compressed (concentrated) gas has a greater free energy than the more expanded (less concentrated) gas, and the difference is given by Equation 3.1.

Equation 3.1 applies to materials in liquid solutions (at dilute concentrations) as well as to gases. When, for example, gaseous oxygen and water come to equilibrium with one another, oxygen is present both in the air and the water, and the oxygen concentration in the water is proportional (by Henry's law) to the pressure of oxygen in the gaseous phase. Since equilibrium requires that there is no effective gradient between gaseous and liquid phases, the free energy of oxygen in the gas is the same as that in the water. Hence the relation of free energy to concentration for the gaseous phase also applies in the liquid. The effect of concentration on free energy affects all chemical and biochemical reactions.

Note that it is the ratio of concentrations (or the reciprocal ratio of volumes) that determines how much free energy has been lost by a gas in expansion. If 1 mole of gas expands from 1 liter to 10 liters, the free energy loss is the same as it would be in expanding from 10 liters to 100 liters. The free energy change is proportional to the absolute temperature. At a standard laboratory temperature of 25°C (298° Kelvin), and given a gas constant, R, of 1.987 calories/degree-mole, a 10-fold decrease in concentration for 1 mole of gas (or solute) would involve a decrease in free energy of:

$$RT\ln\left(\frac{C_1}{C_2}\right) = (1.987)(298.15)\ln\left(\frac{10}{1}\right) = (1.987)(298.15)(2.303) =$$

1364 calories = 1.364 kcal.

At body temperature (37°C = 310.15° Kelvin), the same decrease in concentration would decrease the free energy by 1,419 calories = 1.419 kilocalories.

CAN AN ISOLATED HOMOGENEOUS OBJECT CONTAIN FREE ENERGY?

It is common in both everyday and scientific language to attribute energy to an isolated object or substance. A barrel of oil or an ounce of sugar or a cord of firewood is said to contain energy. Strictly speaking, none of these things in isolation has any free energy (energy that can do work) unless it has access to oxygen or some other chemical with which it can react. Free energy always involves an interaction, either between different entities or between regions containing different concentrations of the same entity. Just as all forces involve interactions between bodies, so all free energies do as well. One might argue that a photon of light contains isolated energy, or a subatomic particle by virtue of its mass, but in neither case will free energy be manifest unless there is an interaction between the photon or particle and something else. Even kinetic energy can only be manifest in reference to another body, and bound energy involves a plurality of microscopic particles.

Nevertheless, it is eminently useful to think and talk as if free energy can be isolated and ascribed to distinct entities: a gram of sugar is said to contain approximately 4 kilocalories of energy, a gram of fat approximately 9 kilocalories, and so forth. A mole of gas confined to 1 liter at 298° K has 1.363 calories more free energy than a mole of gas in 10 liters. (Strictly speaking, neither of them has any free energy unless they have opportunity to expand.) In bioenergetics, we speak of the high-energy phosphate bonds of adenosine triphosphate (ATP), having in mind that a reaction between ATP and water will yield more free energy than some other reactions we can think of.

Tables of data for chemical compounds (Chapter 7) may seem to ascribe energy values to individual substances, but these tables refer to processes that the compound has undergone or could undergo, listing, for example, the free energy of formation of the compound from its elements, or the free energy for combustion, or some other specific reaction of the compound. It is important to remember that all free energy (energy that can do work) involves an interactive process.

Some isolated objects can be homogeneous and yet unstable and hence be able to change spontaneously to a more stable configuration, potentially doing work in the process. Super-cooled water may suddenly freeze and expand, pushing on anything in its environment. Pure tin at low temperatures can spontaneously change from an electrically conductive metallic form to a nonconductive powdery form, expanding in the process.[1] And an explosive chemical such as nitroglycerine contains the oxygen within itself to spontaneously combust with great heat and expansion of gases. With radioactive isotopes, individual atoms are unstable, and their nuclei can spontaneously and without external provocation create a different isotope of the same element or even a different element, as when a neutron within the nucleus of radioactive carbon-14 is converted to a proton, forming ordinary nitrogen-14 with the emission of a high-speed electron (β ray). All these extraordinary cases, however, obey the general rule that free energy is expressed in a process of change and not in a static state.

FREE ENERGY IN EQUILIBRIA OF
OPPOSING GRADIENTS

An equilibrium is a stable state with no tendency to change. It might be the end point of a gradient that has fully discharged,

---

1. This unusual transformation of tin may have contributed to the tragic end in 1912 of the Scott Antarctic expedition, which, on its return trek from the South Pole, found that much of the kerosene it cached on the way in had been lost from leaky cans. The cans had been soldered with tin, which may have deteriorated in the cold.

so that everything is homogeneous and all energy is equipotential (bound). But an equilibrium can also result from a balance of opposing gradients, so that while each gradient alone would have free energy, the two together do not. Such situations are analogous to two flashlight batteries wired positive end to positive end and negative to negative, except that they usually involve two different types of gradients. We will consider three examples. In each case, one of the gradients is a concentration gradient. The opposing gradient is (1) a pressure gradient (osmotic pressure), (2) an electrical gradient (membrane potentials), or (3) a gravitational gradient (sedimentation).

Osmotic Pressure

At some time in the dawn of agriculture, a shepherd probably put a bladder of fresh milk into a mountain brook to keep it cool and was astonished to see it swell and burst open. That was osmotic pressure in action, the same pressure that helps to drive sap up a tree in spring. Like other substances, water has a greater free energy when it is more concentrated, and it tends to diffuse from high concentrations to low. When other substances such as proteins, sugars, and salts are mixed with water (as in milk), the water's concentration is decreased and so is its free energy. If such a solution is placed in contact with pure water, pure water will diffuse into the region containing solutes, and solutes will diffuse toward the pure water. But suppose that a membrane (such as a bladder) is permeable to the water but not to the solutes. Then water will diffuse into the bag, but the solutes are prevented from diffusing out. The bag will swell—unless pressure is exerted on it to keep it from swelling. The pressure needed to keep it from swelling is the osmotic pressure. The movement of water into the bag is called *osmosis*. It is osmosis that causes osmotic pressure, not the other way around.

Consider a cylinder with compartments A containing pure water and B containing water plus solute. The two regions are separated by a rigid membrane, which is permeable to water but not to the solute. At the end of compartment B is a plunger to which a pressure, Π, can be applied—a pressure that is felt by compartment B, but not (because of the rigid membrane) by compartment A. As pure water passes by osmosis down its concentration gradient from A to B, it will start to make B swell and drive the plunger to the right; but, if the plunger is held firmly, a pressure will develop in B instead, which creates a pressure gradient between B and A. Very quickly, the free energy developed in the pressure gradient between B and A balances the free energy of the concentration gradient between A and B, and an equilibrium is established of opposing gradients. The equilibrium is dynamic in the sense that individual water molecules from A still pass into B, but molecules from B pass at an equal rate into A, so the net flux of water is zero.

At osmotic equilibrium, water in A has the same free energy as in B. If this were not so, then a net free energy gradient would exist, and water would pass one way or the other. The free energy involved in moving 1 mole of water across the pressure gradient from B to A is the product of the pressure excess from B to A (which is the osmotic pressure, Π) and the volume, v, of a mole of water:

$$\text{Free energy of pressure gradient} = \Pi v$$

The free energy involved in moving 1 mole of water across the concentration gradient from A to B is $RT \ln\left(\dfrac{C_A}{C_B}\right)$, where C_A and C_B are the concentrations of water in compartments A and B. These concentrations can be expressed as mole fractions—the number of moles of water divided by total moles of water plus solute.

Where n_1 is the number of moles of water and n_2 the number of moles of solute: $C_A = \left(\dfrac{n_1}{n_1} \right) = 1$ and $C_B = \left(\dfrac{n_1}{n_1 + n_2} \right)$.

Then

$$\frac{C_A}{C_B} = \left(\frac{1}{\dfrac{n_1}{n_1 + n_2}} \right) = \frac{n_1 + n_2}{n_1} = 1 + \frac{n_2}{n_1}$$

Setting the free energy of the pressure gradient equal to that of the concentration gradient:

$$\Pi v = RT \ln \frac{C_A}{C_B} = RT \ln \left(1 + \frac{n_2}{n_1} \right) = RT \left(\frac{n_2}{n_1} \right).$$

The simplification in the last step is justified for a dilute solution, where n_2 is small compared to n_1, so that n_2/n_1 is a small number, for $\ln(1 + x) = x$ when x is small. Solving for the osmotic pressure:

$$\Pi = \left(\frac{n_2}{n_1 v} \right) RT = \frac{n_2}{V} RT = C_S RT$$

(The quantity $n_1 v$ is the total volume, V, of compartment B, which contains n_2 moles of solute. The concentration of solute, C_S, in compartment B is n_2/V.)

This equation for the osmotic pressure of dilute solutions was derived purely from free energy analysis by recognizing that the opposing energy gradients of pressure and concentration must be equal at osmotic equilibrium. It does not inform us about the actual mechanics of how the osmotic pressure is generated, but it is striking that it is analogous to the ideal gas law. The osmotic pressure is proportional to the absolute temperature just as is the pressure of an ideal gas, which suggests that it is due to a kinetic impact of the molecules (either solute or water) on the walls of the chamber. It requires, however, a membrane permeable to water but not to solute and the presence of a more dilute solution on the other side of the membrane. There is no osmotic pressure in a bottle of solution sitting on a shelf! The osmotic pressure is also proportional to the concentration of the solute in the same way that the pressure of an ideal gas is proportional to the concentration of the gas.

For a 1-molar solution of impermeable solute, the osmotic pressure generated at 0°C (273°K) is approximately:

$$\Pi = C_S RT = \left(1\frac{mole}{liter}\right)\left(.082\frac{liter-atmos}{mole-\deg}\right)(273\deg)$$
$$= 22.4\,\text{atmospheres}$$

Note that 1 mole of an ideal gas confined to 1 liter at 0°C would also generate a pressure of 22.4 atmospheres. Even a .05-molar solution can generate at 0°C an osmotic pressure of 1.12 atmospheres, enough pressure to support a column of water more than 35 feet high.

Membrane Potentials

Just as a selectively permeable membrane can be involved in the generation of an osmotic pressure if a concentration gradient exists across the membrane, so it can generate an electrical potential if the concentration gradients involve ions of opposite charge, one of which can penetrate the membrane while the other cannot. In this case, the equilibrium reached is one of opposing gradients of concentration and electrical potential.

Suppose we have a membrane between two chambers. Chamber 1 has a high concentration of K^+ and Cl^- ions, while chamber 2 has a low concentration of these ions. If the membrane is permeable to both ions, both will diffuse through the membrane, and eventually an equilibrium will be reached in which the concentrations of K^+ and Cl^- are uniform throughout both chambers, with no gradients or free energy to cause any further change. If, however, the membrane is permeable to K^+ but not to Cl^-, the diffusion of K^+ down its concentration gradient through the membrane from chamber 1 to chamber 2 results in an electrical imbalance, the membrane becoming positively charged in chamber 2 and negatively charged in chamber 1. The positive charge in chamber 2 opposes the further passage of K^+ through the membrane, and an equilibrium of opposing gradients results in which the free energy of the concentration gradient is balanced by that of the electrical gradient. For 1 mole of K^+ ion, the free energy of the electrical gradient is given by the total charge of the ions, which is the Faraday charge, $\mathcal{F}$ = 96,500 coulombs, times the electrical potential, $\mathcal{E}$, across the membrane: electrical free energy = $\mathcal{F}\mathcal{E}$. Balancing this electrical gradient against the concentration gradient gives:

$$\mathcal{F}\mathcal{E} = RT \ln \frac{C_1}{C_2}$$

Solving for the electrical potential gives the Nernst equation, named for Walther Nernst (1864–1941):

$$\mathcal{E} = \frac{RT}{\mathcal{F}} \ln \frac{C_1}{C_2}$$

A typical animal cell has a much higher concentration of K^+ on the inside of the cell than on the outside. For a rat muscle cell, the concentration is about 152 millimolar on the inside and 6 millimolar on the outside. Putting these figures into the Nernst equation, with T = 37°C (310°K) gives:

$$E = \frac{RT}{F} \ln \frac{C_1}{C_2} = \frac{\left(8.3144 \frac{\text{volt}-\text{coulombs}}{\text{deg}}\right)(310 \text{ deg})}{96,500 \text{ coulombs}} \ln \frac{152}{6}$$

$$= 0.086 \text{ volts} = 86 \text{ millivolts}$$

Although the resting cell membrane potential in a muscle or nerve cell is the result of more than just the potassium ion concentration gradient, the K^+ gradient gives a surprisingly good approximation (via the Nernst equation) for the membrane potential. When a nerve or muscle cell is stimulated, the membrane permeabilities change, and Na^+, whose concentration is high on the outside of the cell, rushes in, momentarily reversing the membrane potential, so that the inside of the cell becomes positive to the outside during the nerve (or muscle) impulse. The lives of all our cells depend on these couplings of concentration and electrical gradients across the cell membranes.

Sedimentation Equilibrium and the Earth's Atmosphere

The Earth's gravity is constantly pulling the air around us down. If it did not, we would not be here to talk about it for our atmosphere would long ago have been lost to outer space. But as gravity pulls all the molecules of nitrogen and oxygen down, their thermal energy keeps them continually moving about and prevents them from all settling on or near the ground. A concentration gradient results, with the atmosphere more concentrated at sea level and gradually diminishing toward outer space. An equilibrium results from a balance between the gravitational gradient pulling the air down and the concentration gradient pushing air, by diffusion, upward. The free energies involved define the dimensions of the concentration gradient and, with it, the elevations at which people can live.

Consider the gravitational and diffusive energies involved for 1 mole of air (average molecular weight, M = 28.8 grams). The gravitational free energy for this gas falling from a height, h, is:

$$\text{gravitational free energy} = Mgh$$

The free energy for diffusion across a concentration gradient, from high pressure, P_0, at sea level, to a lower pressure, P, at the same height, h, is:

$$\text{concentration free energy} = RT\ln\frac{P_0}{P}$$

Equating these two free energies gives:

$$RT\ln\frac{P_0}{P} = Mgh$$

$$\ln\frac{P_0}{P} = \frac{Mgh}{RT} \text{ or } \ln\frac{P}{P_0} = -\frac{Mgh}{RT}$$

$$\frac{P}{P_0} = e^{-\left(\frac{Mgh}{RT}\right)}$$

$$P = P_0 e^{-\left(\frac{Mgh}{RT}\right)}$$

The pressure of the air decreases exponentially from its sea level value, P_0, as we ascend to higher elevations, the exponent being Mgh/RT, which is the ratio of the gravitational and thermal (diffusive) energies. The equation assumes a constant temperature, which is an approximation, since the actual atmosphere usually cools with elevation. It also combines nitrogen (80%) and oxygen (20%) together, averaging their molecular weights into one (the gases are kept mixed by convection currents in the atmosphere). At what elevation, according to this analysis, should atmospheric pressure be one-half that at sea level, assuming a constant temperature of 10°C?

$$h = \frac{RT}{Mg}\ln\left(\frac{P_0}{P}\right) = \frac{RT}{Mg}\ln(2) = \frac{(8.3144)(283)}{(28.8 \times 10^{-3})(9.8)}(0.6931)$$
$$= 5779 \text{ meters} = 18,959 \text{ feet}$$

The sedimentation of the earth's atmosphere involves molecules of very small molecular weight, hence the distance calculated above, which is inversely proportional to the particle mass, is very large. For larger particles, or for particles subjected to a greatly amplified force in a centrifuge, the distances for notable changes in concentration are much smaller.

CYCLES OF CHANGE ARE IMPOSSIBLE WITHOUT AN EXTERNAL ENERGY SOURCE

All change is down a gradient of some sort. If B is down-gradient from A, and C is down-gradient from B, then A cannot be down-gradient from C. A river of change has to flow downhill. It cannot flow in a circle—unless there is a pump somewhere—for all points around a circle cannot be downhill from each other! Nothing could be more obvious, yet the point is sometimes overlooked. I studied calculus in college from a very fine book that I still treasure. But near the end of the book it describes the differential equations involved when a radioisotope A transforms to radioisotope B, which transforms to radioisotope C, which transforms back to A, and it makes the point that in such a system of equations the sum of the amounts of A, B, and C remains constant. It is fine mathematics, but it is physically impossible—such a thing cannot happen in the natural world. I had a friend who proposed a theoretical model for the transport of oxygen in a network of arterial vessels. His scheme had oxygen diffusing from A to B and on to C and then back to A. The mathematical equations were fine, but, again, the scheme was physically unreal since diffusion must always be down a concentration gradient, and we cannot have any kind of gradient that is continuously downhill in a circle.

Nature has all sorts of circular changes, processes that fold back upon themselves. The river running across the land takes water to the sea, water evaporates from the sea, and falls upon the land as rain and snow, to drain into the river. But this perpetual motion is driven by an outside energy source, the radiant energy from the sun. Life is full of cyclical reactions. Every enzyme reaction involves at least two steps in which a free enzyme molecule joins with its substrate and then releases the transformed substrate as end product while the enzyme cycles back to its original state. The two steps are each down-gradient because of the overall reaction converting the substrate to end product. Multienzyme cycles such as the Krebs cycle are ingeniously driven by gradients originating outside the cycle, deriving from the metabolism of nutrients supplied to the cell.

Chapter 4

Thermal Energy, Temperature, and Entropy

THERMAL ENERGY (HEAT)

The recognition of equivalence between mechanical work and heat gave rise to the concept of energy in the 1840s. But while mechanical work was measured as a product of two factors—a force and a length—and expressed as a compound unit, such as the foot-pound, heat had long been seen as a single entity and was measured by what appeared to be a simple unit, such as the calorie (the heat required to raise 1 gram of water by $1°C$) or the British thermal unit (BTU; the heat required to raise 1 pound of water by $1°F$). Heat was commonly regarded as a substance, a very subtle fluid, which joined with more tangible substances, making them warm and causing their fine particles or atoms to repel one another and hence to expand outward. Under the French name of "calorique," heat was listed by Lavoisier in his *Elements of Chemistry* as one of the "simple substances," or elements.

There were two problems with regarding heat (or caloric) as a substance. The first was that it could not be weighed. A cold body did not gain any detectable weight upon heating, as Benjamin Thomson (Count Rumford) found with a balance that could detect one part in a million. If heat—that noun that makes something hot—were a substance, it was so light that the finest chemical balance could not record it. The second problem was that heat did not appear to be conserved. It could be produced from mechanical work, as Rumford showed in the boring of cannon and as Joule measured with precision in his experiments, or it could be used up in the production of mechanical work, as Rudolf Clausius saw in his analysis of the steam engine. Heat came to be regarded, as Rumford put it, as "a kind of motion," a motion of the finest particles of which a body is made, particles too fine to be seen with the unaided eye.

Nevertheless, although transformed from a subtle fluid to a kind of motion and to a form of energy, the term "heat" retained a unitary, substantive feeling to it in both everyday and scientific usage, with the result that heat as free energy was not at first split into intensive and extensive factors. The same term, *heat*, has been used to denote the "thing" that flows across a temperature gradient, as well as the energy that accompanies that flow. In electricity, we distinguish between the electric charge that moves across the voltage gradient and the electrical energy that is the product of the charge and voltage. We would not usually speak of volt-coulombs (joules) flowing across a voltage difference, but rather of coulombs of charge (or amps, which are coulombs per second) flowing across the voltage. We distinguish between mass moving across a gravitational potential and the gravitational energy that is the product of the mass and that potential. We would not speak of foot-pounds falling across so many feet of height, but of pounds falling that distance. We certainly distinguish between the mass of reagents involved in a chemical reaction and the energy of the reaction. It is important to make this distinction in thermal systems as well.

Because the noun "heat" has been used ambiguously for both thermal energy and the stuff (entropy) that flows across a temperature gradient, I will usually use "thermal energy" for the first meaning and "entropy" for the second. In chemical energetics, the word "heat" is so deeply ensconced that it would cause confusion to try to avoid it, but when I use "heat" I will always mean energy rather than entropy. The word "heat" is also a verb, meaning to warm, and many chemists regard heat not as something which exists by itself but specifically as one of two forms of energy transfer, the other being work. Other chemists do not limit themselves to this usage, but employ the word more broadly.

Thermal energy is sometimes portrayed as being different from other forms of energy. One will hear that mechanical energy can always be converted in totality to thermal energy, but thermal energy can only be converted in part, or not at all, to mechanical energy. This distinction is the result of two factors. First, thermal energy is the repository of most of the equipotential (bound) energy of our experience (we ignore the bound energy, for example, of mass itself). The discharge of mechanical, electrical, chemical, and other free energy gradients usually is not fully compensated by the recharging of some other gradient and produces entropy (warmth) that results either in an increase of temperature or volume or likely both. Thermal energy is regarded as an equipotential sink into which other forms of energy drain. Second, this reservoir of bound, equipotential energy raises the temperature of the world we live in to well above the absolute zero that constitutes the zero potential for thermal free energy. In other words, much of the thermal energy we experience is bound, and that which is free is associated with a gradient that is truncated by the fact that we live in a warm world.

The truncation of the thermal free energy gradient (the fact that we experience in ordinary life only a small part of its total gradient) can be illustrated by a simple example. Suppose we are living at an ambient temperature of T degrees, and that 1 gram of

water (A), warmed to a temperature of $T + 5°C$, is placed in contact with 1 gram of water (B), chilled to a temperature of $T - 5°C$,
both samples of water being insulated from further exchanges
with their environment. Thermal energy leaves A and enters B,
and when the two have reached thermal equilibrium, they are
both at temperature T. The amount of thermal energy that has
passed from A to B in achieving this equilibrium is evidently 5
calories, as that is the amount required to cool A from $T + 5$ to T
and to warm B from $T - 5$ to T. At the beginning of the thermal
transfer, the extensive carrier (entropy) was falling across a gradient of 10° degrees, but, at the end, its gradient was approaching
zero, so the average gradient during the thermal transfer was 5°.
However, the total thermal gradient that the entropy could fall
across if we could have a sink at absolute zero would be $T°$ rather
than 5°. If T is $27°C \left(300°K\right)$, we have made use of only 5 parts
in 300 of the entire thermal gradient. Entropy falling across 5
degrees can do only 5/300 of the work that it could do if allowed
to fall the entire 300 degrees, and, in falling those 5 degrees, it
has given up only 5/300 of its total possible free energy. Thus, the
transfer of 5 calories of thermal energy from A to B involves a loss
not of 5 calories of free energy from the system, but a loss only
of $(5)(5/300) = 25/300 = .0833$ calories of free energy (and
a gain of an equal amount of bound energy). But it also means
that if the thermal transfer had been completely coupled with a
production of mechanical work, we could only get .0833 calories
of work from 5 calories of thermal transfer. If the same thermal
transfer were done at $200°$ ambient instead of $300°$ (it would require ice rather than liquid water), we could hope to get 0.125
calories of work from the 5 calories of thermal transfer. The realization that thermal energy transferred at higher temperature
is worth less in mechanical action than at lower temperature led
Clausius, in his analysis of the steam engine, to divide thermal

energy by a function of temperature and to refer to this quotient as an "equivalence value" of the energy, which he later decided to call "entropy."

Free thermal energy is exactly analogous to any other free energy: it involves a gradient with a current of "something" flowing across it. The gradient for thermal energy is a temperature difference. The something that flows is entropy. When you put a kettle of water on the hot stove, it is entropy that flows from the stove to the water and raises the temperature of the water. The sum of the increments of entropy times the temperature for each increment is the thermal energy (or heat) received by the kettle. The thermal energy is measured in calories, the temperature in degrees, and the entropy in a compound unit of calories-deg^{-1}, often referred to as "entropy units." No means is yet at hand for defining a unit of entropy by a single measurement, in the way that a unit of charge or mass can be defined, for while massive bodies exert a gravitational force on one another, and charged bodies exert an electric force, the entropy of bodies does not exert a measurable force at a macroscopic level. The entropy unit is a composite of thermal energy divided by temperature, and thus entropy has the dimensions of calories divided by degrees, or in mechanical terms of $ML^2T^{-2}\Theta^{-1}$, where Θ (theta) denotes the dimension of temperature as measured by some sort of thermometer.

TEMPERATURE

Sensory Judgments of Warmth and Cold

Hot and cold are attributes of our world that are continually with us in our lives, monitored by peripheral nerve endings in our skin (Ruffini nerve endings for warmth and Krause end bulbs for cold) and by nerve cells in the hypothalamus of our brain, tuned to protect our central body temperature by controlling blood flow, regulating the metabolic rate of brown fat, and inducing shivering against cold or sweating against warmth. In ancient natural philosophy, hot and cold, along with wet and dry, were the fundamental pairs of opposing properties expressed in the four elements of nature and in the four humors of the human body: fire was hot and dry, for example, and water cold and wet.

It is not directly evident in our everyday lives that there is a continuous scale of temperature, nor whether, if there is a scale, it is one of warmth or of cold. Both hot and cold seem to be, in different contexts, equally positive and real qualities. Just ask the cook who inadvertently touched the oven grill or the winning sports coach who has been doused with ice water. If there is a continuous scale for measuring warmth or coolness, should it start at the warm end and measure degrees of cold, or start at the cold end and measure degrees of warmth, or perhaps start in a zone of neutrality (like an acid–base scale) and measure each way? (When Anders Celsius first devised his thermometer, he put $0°$ at the boiling point of water, and $100°$ at the freezing point, thus in effect measuring degrees of cold rather than warmth.) Moreover, our senses play tricks on us. They respond to rates of change of warmth or coldness as well as to warmth or coldness itself. A cold metal seat feels colder than a wood seat right next to it because it conducts warmth away from us faster than the wood. Our senses also accommodate to stimuli and are influenced by their previous history. A hand plunged into a bucket of water at room temperature may feel cool if it had previously been in warm water and warm if it had previously been in ice water.

Thermometers

Temperature could only become an objective measurement when it was observed that outside objects respond in a predictable way to what our senses perceive as warmth or coldness. Today, many thermal-sensitive elements are built into electrical devices, making use of the effect that temperature has upon electrical properties. A *thermistor* detects temperature changes by changes in an electrical resistance—a higher temperature causes a greater resistance in the element. A *thermocouple* makes use of the dependence on temperature of the electrical potential at a junction of two dissimilar metals. But the earliest thermometers were based on the expansion and contraction of materials when temperature goes up or down. Materials usually expand when heated and contract when cooled, and the change in volume of a liquid or gas was the basis for the first thermometers. Today, the expansion or contraction of metals is also used in thermometers and thermostats: a bimetallic plate will bend with temperature changes if it is made of metals having different thermal expansion coefficients.

For a moderate change in warmth, the change in volume of any fluid is small, so that it is necessary in making a thermometer to magnify the changes by connecting a relatively large reservoir of fluid to an expansion tube of small diameter. The excursion of the fluid meniscus along the fine tube then becomes significant for small changes in the volume of the fluid. Thermometers designed for high precision have very large bulbs and very fine capillary tubes—but also have the limitation that they can be used over only a small range of temperatures. The cross-sectional area of the fine tube needs to be uniform in order for the linear excursion of the meniscus of the fluid to be exactly proportional to the change in volume of the fluid. Every fluid thermometer is likely to have a slightly different behavior because of differences in dimensions. The behavior of the thermometer also depends on the walls of the reservoir and tube that contain the fluid, as these walls will also expand or contract as they are warmed or cooled. The walls will actually respond to changes in environmental temperature faster than the liquid because the liquid can only experience a change in temperature from the walls. If a mercury thermometer is suddenly plunged into a warm bath, the mercury column may momentarily fall a little (as the walls expand) before it begins to rise.

Anyone can engrave markings on the tube of a thermometer and declare that they represent a scale of temperature, but the scale would be entirely arbitrary: that is, it would not agree with anyone else's thermometer unless the two were compared side by side under various common conditions of warmth and a table of corrections made for one scale or the other. Thermometer scales can be compared to one another without bringing the thermometers together if a universal reference state can be agreed upon, like sea level as a reference point for elevation. In the 17th century, Edmund Halley suggested that the Grottoes deep under the Observatory at Paris have an unchanging temperature that could be used as a reference point—a strange suggestion coming from an Englishman, for who wants to go all the way to Paris to calibrate his thermometer? The freezing point and boiling point of water are now taken as fixed thermal points that everyone can use for calibrating thermometers, but they have complications because they are affected not only by atmospheric pressure and solute contamination, but also by super-cooling and super-heating effects. When water freezes, it usually super-cools before the first crystals of ice form, and then suddenly the temperature leaps up to a plateau that remains stable over time. It is quite startling to see this on a precision thermometer. When water boils, there is a range of temperatures within the water itself, but the steam above gives more consistent readings.

A precise (but inconvenient) reference point is the triple point of water, which is the temperature (and pressure) at which all three phases of water—ice, water, and water vapor—are at equilibrium with one another. The triple-point temperature is just above the normal freezing point at $.01°C$ $(273.16°K)$, while the triple-point pressure is 4.588 torr (mm Hg). That there is only one temperature and pressure at which the three phases of water can be at equilibrium with one another is explained by the phase rule of Josiah Willard Gibbs (1839–1903):

$$f = 2 + c - p$$

where f is the number of factors that can be independently varied in a system (the "degrees of freedom"), c is the number of different chemical components, and p is the number of phases (1, 2, or 3). If we start with pure liquid water, we have one component $(c = 1)$ and one phase $(p = 1)$, and we can independently vary two "state functions" such as temperature and pressure. The degrees of freedom are two, which is the baseline number for any pure substance residing in one phase. Each time an additional component is added (like salt to water), we gain one degree of freedom in being able to vary the concentration of that component. But each time an additional phase is added, we lose one degree of freedom because the free energy of one phase must equal that of the next, and their respective free energies are affected in different ways by changes in temperature and pressure. Thus, if we have water and ice together, we lose a degree of freedom because changes in temperature and pressure affect the two phases in different ways. If pressure is increased above atmospheric, the ice (which is less dense) tends to collapse to water (which happens under the blade of an ice skate or under the weight of a glacier), so that, to maintain equilibrium, the temperature must be decreased to favor the formation of ice. Any change in either pressure or temperature has to be accompanied by an opposite change in the other: the two factors cannot be varied independently if equilibrium is to be maintained. If a water vapor phase is added to the ice and water, then no freedom is left: the temperature must be raised to .01°C, and the pressure must be greatly reduced from atmospheric, down to the vapor pressure that water has at .01°C, which is 4.588 torr (mm Hg).

Although water can be used to make a barometer, such as the old "water glass" that hangs in my house (which is actually affected by temperature as well as barometric pressure), water is not versatile in a thermometer because of the danger of its freezing and breaking its container. A water-filled thermometer would not be of much use below 8°C because water reaches a maximum density at 4°C and then begins to expand before it freezes. A water thermometer would read nearly the same at 8°C as at 0°C, and if it read "8" one would not know that it did not mean "0" (unless one had been watching its prior readings). Likewise, the marking on the thermometer for 7° would also represent 1°, 6° would share its marking with 2° and 5° with 3°. Does anyone want to buy a water-filled thermometer? Ethanol, on the other hand, does not freeze until the temperature reaches – 114.5°C, and it does not expand before freezing.

The first mercury thermometer was made by Daniel Fahrenheit in 1714. The mercury thermometer outside my northern Minnesota house has no markings below −40°C. Why? Because mercury freezes at −38.87°C. Thankfully, if a winter night goes below that temperature, the thermometer will not break as mercury does not expand on freezing. When Fahrenheit calibrated his mercury thermometer (and thereby invented the Fahrenheit scale), he set his 0° mark at the temperature of an ice-water bath saturated with ammonium chloride salt. For his next reference mark, he used an ice-water mixture without the salt and decided to call this mark 32°. He could have picked any number, and 32 may seem a strange choice, but nonmetric carpenters' rulers are

often divided into 32^{nds} of an inch. As a third reference temperature, Fahrenheit used the human underarm temperature, which he
judged to be $96°$, three times farther from his zero point than was
the freezing point of water. It was soon decided that the boiling
point of water is 180 Fahrenheit degrees above the freezing point
and hence is $212°$.

In 1742, Anders Celsius of Sweden created a scale of $100°$ between the freezing point and boiling point of water. Initially, the
boiling point was $0°$ and the freezing point was $100°$, so that the
scale measured degrees of cold rather than of warmth, but this
was soon reversed at the suggestion of the French, who called it
the *centigrade scale*. Centigrade and Celsius now mean the same
thing. The scale has $100°$ for the same range (freezing point to
boiling point of water) covered by $180°$ of the Fahrenheit scale, so
that the centigrade degree equals 1.8 Fahrenheit degrees. To convert a centigrade reading to Fahrenheit:

$$F = \frac{9}{5}C + 32$$

To convert a Fahrenheit reading to centigrade:

$$C = (F - 32)\left(\frac{5}{9}\right)$$

A thermometer checked against an ice-water bath and a
steam-water bath can be marked with two reference points, 0 and
100. By measuring the distance along the tube between the two
reference points and dividing by 100, 99 new marks can be made
at equal distances and labeled 1 through 99. While two different
thermometers made in this fashion should read the same at 0
and 100, they will not necessarily read the same at 50 because of
slight nonuniformities in the bores of their tubes. Nor have we

any assurance that 50 on either one of them represents a "temperature" exactly halfway between 0 and 100. In trying to measure temperature, we are really measuring the volume of a liquid. Assuming that our thermometer tube is uniform and our volume measurements are right, are we really happy using a volume measurement to tell us the temperature? To say that the equally spaced marks along the tube represent equal "degrees" of temperature, we must assume that the volume of the liquid changes linearly with temperature: that the liquid, starting with a volume, V, at a reference temperature, has a constant coefficient of expansion with temperature, that $\dfrac{dV}{dT} =$ constant. We cannot expect this to be the case for more than a limited range of temperature. And how can we know that a liquid changes its volume linearly with temperature unless we have some other independent means of measuring temperature? No matter how accustomed we become to gazing at a thermometer and supposing that the length of its column is an exact representation of a true temperature scale, we have to admit that we are jumping to comforting conclusions.

The possible errors in temperature readings can be explored by comparing the behavior of thermometers filled with different liquids; the differences in readings for the range $0-100^{\circ}$C are usually small. But the temperature scales that agree most closely are those based on the volumes or pressures of gases, especially those such as H_2 that liquefy only at very low temperatures and behave very nearly like ideal gases. By their similar behavior, such gases give reassurance that we are really measuring something "absolute" rather than something relative only to particular substances. Starting at 0°C, they all expand or contract by about one part in 273.15 of their original volume for each Centigrade degree change in temperature. A plot of volume at constant pressure against temperature gives a constant slope, with an extrapolated intercept at $-273.15°C$ where the volume is zero. A zero volume is of course impossible for real matter, and all real gases liquefy before a temperature of $-273.15°C$ is reached, but the behavior of gases suggests the existence of an absolute zero of temperature at $-273.15°C$ (or $-459.67°F$). The Kelvin scale (named for William Thomson, Lord Kelvin) uses Centigrade degrees but with its zero at -273.15, so that a reading on the Kelvin scale is 273.15 degrees higher than on the Celsius (Centigrade) scale. The Rankine scale (named for William Rankine) starts at absolute zero but uses Fahrenheit degrees, so that a reading on the Rankine scale is 459.67 degrees higher than a reading on the Fahrenheit scale. If a winter day seems cold when the thermometer reads $0°F$, we can be grateful that we are 459.67 Fahrenheit degrees warmer than absolute zero.

Since there are 1.8 Fahrenheit degrees in a Celsius degree, a reading on the Rankine scale is 1.8 times that on the Kelvin scale. Thus:

$$K = C + 273.15$$
$$R = F + 459.67$$
$$R = 1.8K$$

where K, C, R, and F signify readings on the Kelvin, Rankine, Celsius, and Fahrenheit scales, respectively.

The ideal gas law predicts the behavior of gases using an absolute temperature scale: $PV = nRT$ where T denotes absolute temperature

$$P = \left(\frac{nR}{V}\right)T = kT$$

where the constant k incorporates the constants n, R, and V, assuming that n and V are held constant while P is allowed to vary with T. Pressure increases with the absolute temperature, with a slope of k, and has a zero value when the absolute temperature is zero.

The change of pressure with temperature is linear for an ideal gas no matter what temperature scale is used (the gas does not know what temperature scale we are using), but the slope depends on the size of the degree for the temperature scale, and the pressure intercept depends on where the zero for temperature is chosen to be. In general:

$$P = P_0 + mt$$

where m is the slope, t is the temperature expressed in any temperature scale, and P_0 is the value of P when t is zero. Since the equation must be dimensionally homogeneous, m must have the dimensions of pressure divided by temperature. The expansion coefficient, α, for the gas is the slope normalized by dividing by P_0 so that $\alpha = m / P_0$ or $m = \alpha P_0$. Then:

$$P = P_0 + \alpha P_0 t = P_0 \left(1 + \alpha t\right)$$

The ratio of two pressures at different temperatures is:

$$\frac{P_1}{P_2} = \frac{P_0(1 + \alpha t_1)}{P_0(1 + \alpha t_2)} = \frac{\dfrac{1}{\alpha} + t_1}{\dfrac{1}{\alpha} + t_2}$$

If Centigrade degrees are used, the expansion coefficient, α, is $1/273.15$, so that $1/\alpha$ is 273.15. The ratio of pressures then becomes:

$$\frac{P_1}{P_2} = \frac{273.15 + t_1}{273.15 + t_2}$$

And if then we let $T_1 = 273.15 + t_1$ and $T_2 = 273.15 + t_2$, the ratio of pressures becomes:

$$\frac{P_1}{P_2} = \frac{T_1}{T_2}$$

This illustrates the convenience of using the absolute temperature scale for the description of gases. But the absolute temperature scale, which was discovered empirically through the study of gases, took on a much greater significance when Kelvin gave it a theoretical meaning, suggesting that the temperature scale should be a measure of thermal energy.

Temperature and Thermal Energy

Kelvin proposed, in effect, that a temperature scale should be a true measure of the thermal gradient. For a given amount of entropy falling across the gradient, each degree in the temperature scale should be equivalent to every other degree in terms of the mechanical work that could be obtained from the entropy flow. The degrees do not have to be Celsius degrees—they could be of any size—but they have to represent equal units of work (free energy) available from flow across them. If the temperature scale also starts at an absolute zero, then the difference in any two temperatures can be compared to the difference between the higher temperature and absolute zero to determine what fraction of the total possible thermal free energy is available for the production of mechanical work. This is the meaning of the absolute temperature scale for the dynamics of thermal free energy, for determining what fraction of total thermal energy is free to be coupled to the production of work, to the building of a gradient in a mechanical or other system.

The absolute temperature scale also gives a measure of the equipotential (bound) thermal energy that exists in a system. The ideal gas law, $PV = nRT$, is an expression of this fact for an ideal gas, that the PV energy of the gas is linearly proportional to the absolute temperature, T. The energy of an ideal gas depends only on the absolute temperature: it is a measure of the absolute temperature, just as the absolute temperature is a measure of it. All the energy of an ideal gas is energy of motion—motion of translation, vibration, and rotation—and the total of this energy increases linearly with temperature and expresses itself most obviously in the pressure exerted by the gas on a containing surface. But the internal energy of all substances increases with temperature, and the free energy gradients of chemical systems are dramatically affected by temperature. We have seen already that the free energy of concentration gradients involved in such phenomena as osmotic pressure, membrane potentials, and sedimentation is linearly proportional to absolute temperature.

ENTROPY

When a gradient of free energy discharges, as when an electrical voltage drives current through a wire, either of two things can happen. In the first case, work is done, which means that the discharge of the electrical gradient causes the recharge of some other gradient, such as the lifting of a weight, the concentrating of a chemical component, or the pumping of a fluid volume against a pressure. The free energy of the electrical discharge is not converted to bound energy, but is *compensated*, meaning that it is replaced by free energy of a different form. In the second case, work is not done, but the free energy of the electrical discharge becomes bound, equipotential energy which cannot carry out any further change. The electrical discharge is said to be *uncompensated*. The electrical wire gets hot and warms the room, but no work is done.

In actuality, it does not seem to be possible for the first case to happen without the simultaneous occurrence of some of the second. No discharge of a free energy gradient is compensated 100% in the world we live in. In the ideal world of the mind, we can see it happening, but in the real world of experience and of engineering we always have some sort of friction that makes a pendulum stop swinging or a toy top stop spinning. Losses of free energy always occur. But that is not altogether a bad thing, for without it we could not live, the world could not go forward in historical progression, but would forever hang in a balance of timeless uncertainty. Just as your car cannot go forward without friction between the tires and the road, so life and the cosmos cannot go forward without decreases in free energy. It is the possibility for a decrease in free energy that points the direction of a chemical change in the laboratory and in our living cells. Life would not be possible without it.

When free energy decreases, entropy increases. The two processes go hand in hand, or, better, the two are one process seen from different perspectives. Every spontaneous change (one that occurs without outside energy input) is accompanied by an increase in entropy—an increase in microscopic agitation—that evidences itself in a change in temperature, volume, concentration, or phase. This microscopic agitation causes molecules to spread out if not otherwise constrained—it is a spreading agency. It is also the "stuff" that can flow across a temperature gradient. It is the extensive factor of thermal free energy.

Unlike the words "heat" and "hot" that derive from a common old Teutonic origin, the first a noun and the second an adjective, the word "entropy" was sprung upon the scientific community in

1865 by the brilliant German physicist, Rudolf Clausius. In trying to understand the relationship between temperature differences and the ability of "heat" (thermal energy) to produce mechanical energy in a steam engine, Clausius came to realize that, in view of the new principle of the conservation of energy (that the sum of heat and work was conserved), heat itself could not be conserved if at the same time it produced work. He looked for something that was conserved under ideal conditions in the engine and found it in what he at first called the "equivalence value" of the heat (heat divided by temperature), and later called the "transformational content" (signified by the letter S). In 1865, he was ready for a new name and went to his Greek dictionary, perhaps inspired by Michael Faraday's previous invention (with the help of a Cambridge classicist, William Whewell) of such words as electrode, cathode, anode, electrolyte, cation, and anion. As Clausius wrote in his ninth memoir on the subject:

> We might call S the *transformational content* of the body, just as we termed U its *thermal and ergonal content*. But as I hold it to be better to borrow terms for important magnitudes from the ancient languages, I propose to call the magnitude S the *entropy* of the body, from the Greek word τροπη, *transformation*. I have intentionally formed the word *entropy* so as to be as similar as possible to the word *energy*, for the two magnitudes to be denoted by these words are so nearly allied in their physical meanings, that a certain similarity in designation appears to be desirable.[1]

1. This and the following quotation are from Clausius, Rudolf Julius Emmanuel *The Mechanical Theory of Heat, with its Applications to the Steam-Engine and to the Physical Properties of Bodies.* Translated by John Tyndall. Edited by Thomas Archer Hirst. London: John van Voorst, 1867.

At the conclusion of this ninth (and last) memoir on the steam engine, Clausius used his new word in giving us his famous declaration of the two laws of thermodynamics:

> We may express in the following manner the fundamental laws of the universe which correspond to the two fundamental theorems of the mechanical theory of heat:
> 1. *The energy of the universe is constant.*
> 2. *The entropy of the universe tends to a maximum.*

Clausius's nine memoirs, published in English translation in 1867 under the title *The Mechanical Theory of Heat with Its Applications to the Steam Engine and to the Physical Properties of Bodies*, constitute one of the truly great contributions to the history of physics. But the term "entropy" was not a happy choice for the very reason that Clausius liked it—that it sounded so much like "energy." It is not energy (it does not have the dimensions of energy), but many distinguished scientists have written that it is a degraded or unavailable form of energy. Furthermore, it entered the lexicon of science through a formalistic definition that was difficult to grasp and nearly impossible to visualize. As Hans Fuchs has noted in his pioneering work, *The Dynamics of Heat* (1996):

> It is almost impossible to recognize the true nature of a quantity which is introduced on purely formal grounds and led onto the stage through the back entrance only. If we were to teach basic electricity in a manner analogous to that found in the chapters on thermodynamics of our introductory physics texts, we would never realize that there is a quantity with the properties of electric charge.

The term "entropy" caused confusion from the beginning. In 1868, Peter Guthrie Tait published *Sketch of Thermodynamics*, in which he not only presented entropy as a kind of energy, but completely twisted Clausius's meaning by saying that it was available energy (rather than a factor in unavailable energy). Tait's book was read by his friend James Clerk Maxwell, who was so confused by it that in the first edition (1871) of his own *Theory of Heat* he used the term entropy to mean, at different places in his book, either available energy or unavailable energy. It was by reading the papers of Josiah Willard Gibbs that Maxwell came to realize that entropy was not energy at all, but was the same thing that William Rankine had called the "thermodynamic factor," and Maxwell therefore corrected his presentation in his (posthumous) edition of 1888. Tait, having been scolded by Maxwell for misleading him, corrected the second edition of his own book in 1877. In more recent times, the great biochemist, Albert Lehninger, in the first edition of his classic *Bioenergetics* (1965), wrote "Free energy is thus 'useful' energy, entropy is 'degraded' energy." He corrected himself in the second edition (1971), but the discussion of entropy remained far less satisfactory than the remainder of his illuminating book. In his fine book, *Engines, Energy, and Entropy* (1982), John B. Fenn noted, "We will introduce an enigmatic quantity called *entropy* that has perplexed many generations of students. Just as many generations of teachers have been confounded by attempting to explain what it is." W. F. Luder, in his *A Different Approach to Thermodynamics* (1967), recommended to his reader that "if he finds himself worrying about his inability to understand entropy . . . he should remember that he will not need to understand it . . . human beings do not—and probably cannot, in this life at least—understand anything fundamental. No one understands gravitation; yet few people worry about falling off the earth."

Entropy is often identified in scientific literature, and especially in the common language, as a measure of disorder or disorganization—or as a kind of dark agency that leads to disorder, disorganization, or to death. I have heard life described as a "fight against entropy." Yet without entropy, there would be no life. For one thing, we would all be frozen at absolute zero. Entropy is what warms us. Without entropy, there would be no dynamic state to matter, atoms and molecules would not move about, and there would be no diffusion of matter from one place to another. Every living cell is dependent upon the dispersal of materials by diffusion. Without diffusion, without entropy, oxygen would not enter our blood from the air in our lungs, nor pass to the billions of our living cells from the blood. Without entropy, carbon dioxide would not enter the green leaf of the plant, nor water and minerals enter its roots. Under a warm shower, give thanks for entropy, and let us not consign it to the category of evil demons. And if the existence of entropy is necessary for life, essential, too, is its increase in spontaneous processes, for entropy increase is companion to free energy decrease in pointing the direction of chemical transformation and physical transport in all the orderly activities of living organisms.

THE MEASUREMENT OF ENTROPY

Unlike mass or charge, entropy exerts no force (at a macroscopic level) that can be detected. The measurement of entropy depends on its capacity for conveying thermal energy at a given temperature, and hence its assessment involves the measurements of both thermal energy and temperature. By a curious paradox, thermal energy, which is measured as one "thing" by calorimetric methods, is really a composite of its intensive element, temperature, and extensive element, entropy; but entropy, which is more justly considered one thing, is measured as a composite of thermal energy divided by temperature, with dimensions of energy/temperature, or $ML^2T^{-2}\Theta^{-1}$, where Θ is the dimension of temperature.

It is assumed that, at absolute zero, the entropy of a pure substance is zero, a postulate that is sometimes called the third law of thermodynamics. This certainly makes sense if entropy is regarded as the kinetic movement or agitation of matter. The energy of an ideal gas is proportional to its absolute temperature and is zero at absolute zero, where the molecules have no kinetic energy. For those, following Ludwig Boltzmann, who view entropy as a probability function measuring disorder or indefiniteness, zero entropy at absolute zero temperature also makes sense since, without any agitation, there can be only one stationary state for a pure substance and no uncertainty about which of many microscopic states the individual atoms or molecules may occupy.

Measuring up from absolute zero, entropy accumulates as a body is warmed by an influx of entropy. The entropy influx is measured, however, as thermal energy (heat) divided by temperature. Ideally, the heating of the substance should be carried out "reversibly" so that the heated body is always at equilibrium, and only an infinitesimal difference exists between the temperatures of the body and its entropy source, allowing by the slightest change in conditions the entropy flow to be reversed. This condition is, of course, impossible in practice as it implies an infinitely slow entropy influx. But in theory it is necessary for two reasons: (1) so that the temperature of the heated body is uniform and therefore measurable as one temperature and (2) so that no entropy is created in the heating process (more about this later).

The amount of entropy required to increase the temperature of a body by a given amount is the entropy capacity of the body, but chemists use "heat capacities" rather than entropy capacities, since it is energy rather than entropy that is directly measured in calorimetric studies. The heat capacity is the energy required to warm a body by a given amount and is equal to the entropy capacity multiplied by the temperature. The heat capacity for an ideal gas is constant, being the gas constant, R, and the entropy capacity is inversely proportional to the absolute temperature, being R / T. But for real substances, the heat capacity is not constant except at high temperatures and approaches zero at temperatures close to absolute zero. The heat capacity at constant pressure, C_p is:

$$C_p = \frac{dq}{dT} \quad \text{and thus} \quad dq = C_p dT$$

where q is thermal energy and T is temperature. Integrating between $T = 0$ and any other value for T to find the thermal energy added:

$$\int_0^q dq = q = \int_0^T C_p dT$$

For entropy, S:

$$dS = \frac{dq}{T} \quad \text{if } dq \text{ is reversible}$$

$$\int_0^S dS = S = \int_0^T \frac{C_p dT}{T} = \int_0^T C_p d(\ln T)$$

A plot of $\frac{C_p}{T}$ against T or of C_p against $\ln T$ gives entropy as the area under the curve, but it is breathtaking to think of the years of labor that must have gone into all the heat capacity measurements necessary to give us the tables of entropy data that we have today. For a limited temperature range between T_1 and T_2 over which C_p is taken as constant:

$$\int_{S_1}^{S_2} dS = (S_2 - S_1) = C_p \int_{T_1}^{T_2} \frac{dT}{T} = C_p \ln\left(\frac{T_2}{T_1}\right)$$

Over modest temperature ranges, entropy goes up approximately with the logarithm of the absolute temperature.

With a change of phase from solid to liquid, or from liquid to gas, entropy and its accompanying thermal energy are absorbed without any change in temperature, the energy being used to weaken or break chemical bonds. It has been customary since the time of Joseph Black (1760s) to describe this as "latent heat," however inconsistent this may be with the way "heat" is used in other contexts, and today some writers also refer to "latent entropy." In melting 1 gram of ice, about 80 calories of thermal energy are required even though the temperature remains steady at $0°$C. The entropy goes up by $80/273 = 0.293$ entropy units per gram or 5.27 entropy units per mole without any temperature change. For vaporization of water at $100°$C, about 540 calories per gram are required, which corresponds to $540/373 = 1.45$ entropy units per gram or 26.1 entropy units per mole. The entropy content of a substance depends not only on its temperature, therefore, but also sometimes on the rearrangement of atomic or molecular interactions, or what Rudolf Clausius called the "disgregation" of a substance, phase changes being the most dramatic examples.

Entropy is also a function of the volume or concentration of a substance, which can change without any change of temperature, as in gases, and also in liquids with dissolved solutes or mixtures of solvents. In the previous discussion, using C_p (the heat capacity at constant pressure) disguised the role of volume in contributing to entropy. If we substitute $(C_v + PdV)$ for C_p and consider the case of entropy changes in 1 mole of an ideal gas:

$$dS = \frac{dq}{T} = \frac{C_v dT}{T} + \frac{PdV}{T} = \frac{C_v dT}{T} + \frac{RdV}{V}$$

The last step above makes use of the fact that, in an ideal gas, $\frac{P}{T} = \frac{R}{V}$.

In integrating between temperatures T_1 and T_2, C_v can be put outside the integration since it is constant for an ideal gas:

$$\int_{S_1}^{S_2} dS = (S_2 - S_1) = C_v \int_{T_1}^{T_2} \frac{dT}{T} + R \int_{V_1}^{V_2} \frac{dV}{V} = C_v \ln \frac{T_2}{T_1} + R \ln \frac{V_2}{V_1}$$

Hence, for an ideal gas, entropy increases not only with the logarithm of the temperature, but also with the logarithm of the volume as well. Since volume is inversely related to concentration (for a given amount of substance), the preceding could be written to show explicitly the effect of concentration on entropy:

$$(S_2 - S_1) = \Delta S = C_v \ln \frac{T_2}{T_1} + R \ln \frac{C_1}{C_2}$$

The $R \ln \frac{C_1}{C_2}$ factor may look familiar. Multiplied by the absolute temperature, it becomes $RT \ln \frac{C_1}{C_2}$, which is the free energy of a concentration gradient, such as we met in discussing osmotic pressure, membrane potentials, and sedimentation. The free energy decrease in a concentration gradient, divided by the temperature, is the same as the entropy increase due to a change in concentration.

FREE ENERGY AND ENTROPY
IN THERMAL CONDUCTION

Imagine two cubes of metal, 1 and 2, connected to one another by a metal wire, the whole system being thermally and electrically insulated from its environment. A switch, which we can open or close from outside, lies between cube 1 and the wire, and we can also selectively pass electrical charge or entropy to cube 1 from the outside so as to create a higher electrical or thermal potential in cube 1 as compared to cube 2.

Suppose we electrify 1 and then close the switch, so we have an electrical potential between 1 and 2. Free energy from that gradient might be used to run a tiny electric motor, but instead we let the electric charge pass through the wire uncompensated by any work, so that all the electrical free energy becomes bound energy. The bound energy appears in the form of thermal energy, causing the wire and eventually both cubes to warm a bit. Entropy has been produced in the disappearance of the free energy and has caused the wire and cubes to warm. We have traded electrical free energy for thermal bound energy.

Now we repeat the procedure, except this time we introduce entropy into cube 1 instead of electric charge. The entropy causes the temperature of cube 1 to rise above that of 2, and, when we close the switch, we have a thermal gradient between 1 and 2, with thermal free energy that might theoretically do a little work. But instead we let entropy flow through the wire from 1 to 2, uncompensated by any work, until the temperature of the entire system is uniform. A peculiar question arises: If entropy and bound thermal energy are created in the wire by the uncompensated discharge of an electrical gradient, are they also created by the uncompensated discharge of a thermal gradient?

Can the flow of heat create heat? The conservation of energy says no, and the entire science of calorimetry is founded on the assumption that thermal energy can be passed from one body to another without any gain or loss. The way out of this puzzlement is this: Just as the uncompensated loss of electrical free energy creates entropy, so does the uncompensated loss of thermal free energy. The increase in entropy in each case causes an increase in bound thermal energy, but whereas in the electrical discharge the thermal bound energy has come from electrical free energy, in the thermal discharge the thermal bound energy has come from thermal free energy, so that while thermal energy increases in the electrical discharge, it does not in the thermal discharge. Symbolically, the change of thermal energy, TS, with time, t, can be viewed this way:

$$\frac{d(TS)}{dt} = \frac{TdS}{dt} + \frac{SdT}{dt} = 0$$

The first term on the right is positive because dS is positive (entropy increases), but the second term is negative because dT is negative (the temperature gradient decreases with time). In absolute value, the two terms must be equal, as dictated by the conservation of energy. So the increase in thermal bound energy (the first term) is exactly equal to the decrease in thermal free energy (the second term), and although entropy increases in the thermal case as in the electrical, the thermal flow does not produce net thermal energy as does the electrical flow.

Let us calculate the increase in entropy and loss of free energy for uncompensated thermal conduction; that is, for entropy transfer that proceeds without the accomplishment of any work. Suppose 1 mole of substance at temperature T_1 is brought into contact with one mole of the same substance at a lower temperature T_2. At the moment of contact, a thermal gradient of $T_1 - T_2$ exists, but, as conduction of entropy proceeds, that gradient gradually diminishes until thermal equilibrium is reached. The equilibrium requires that the original free energy gradient has been completely destroyed, so that all the original free energy has been converted to bound energy. The absence of a gradient means that the temperature has become uniform throughout the system (all thermal energy is equipotential). In this particular case, because the amounts and heat capacities of the two substances are equal, the equilibrium temperature, T, is midway between T_1 and T_2, and can be expressed as $T = \dfrac{T_1 + T_2}{2}$ or as $T = T_2 + \dfrac{T_1 - T_2}{2}$. If the system is at constant pressure, and if the heat capacity, C_p, is constant across the temperature range, then the entropy changes undergone by the warmer body 1 and the colder body 2 are:

$$\Delta S_1 = C_p \ln \frac{T}{T_1}$$

$$\Delta S_2 = C_p \ln \frac{T}{T_2}$$

Since the equilibrium temperature, T, is smaller than T_1 but greater than T_2, the right-hand term in the first equation is negative, but, in the second equation, it is positive, meaning that the entropy of the warmer body, A, has decreased (since entropy has flowed from it), while the entropy of the initially cooler body, B, has increased. The interesting point, however, is that the entropy of B has increased more than the entropy of A has decreased, meaning that B has not only received the entropy sent from A, but additional entropy that was created in the process—the entropy produced from the uncompensated loss of the original thermal free energy. This is seen if we take the sum of ΔS_1 and ΔS_2:

$$\Delta S_T = \Delta S_1 + \Delta S_2 = C_p \ln\frac{T}{T_1} + C_p \ln\frac{T}{T_2} = C_p \ln\frac{T^2}{T_1 T_2}$$

The result is positive, since T^2 is necessarily greater than $T_1 T_2$, just as a square is necessarily greater than a rectangle of the same perimeter. For T_1 is $T+x$ and T_2 is $T-x$, where x is $\left(T_1 - T_2\right)/2$. Then, $T_1 T_2 = (T+x)(T-x) = T^2 - x^2$, which is necessarily less than T^2. Thus, entropy has increased by the amount $C_p \ln\dfrac{T^2}{T_1 T_2}$ in the conduction of entropy from A to B. The entropy increase multiplied by the average (equilibrium) temperature T gives the increase in thermal bound energy, which is also the decrease in thermal free energy, ΔG:

$$\Delta G = -T\Delta S = -TC_p \ln\frac{T^2}{T_1 T_2}$$

In going to equilibrium, where the temperature is uniform and a thermal free energy gradient no longer exists, the system has also maximized its entropy. Suppose that the system had stopped a bit short of reaching equilibrium, so that the final temperatures for A and B were $T + \varepsilon$ and $T - \varepsilon$. Then the numerator in the logarithm for the calculation of ΔS_T would be $T^2 - \varepsilon^2$ instead of T^2, thus giving a lesser amount of entropy increase. Thus, in going to a uniform temperature at equilibrium, the system has simultaneously reduced its thermal free energy to zero (because there is no longer any thermal gradient) and increased its entropy to a maximum value.

EQUILIBRIUM AND LE CHATELIER'S PRINCIPLE

An equilibrium is a stable condition that does not change with time. For such a state to exist, there must be no internal energy gradients making the system change, and it must tend to resist any disturbing forces from outside. An example of the requirements for equilibrium can be seen in the way that a chemical double-pan balance is constructed. A simple balance beam placed upon a knife edge, like a children's teeter-totter, may have temporary stability if the weight of the beam is evenly distributed, but as soon as an additional weight is placed on one end (like a single child climbing on the see-saw), the beam will tilt and, in the absence of any other force, will continue to rotate until it slides right off its point of rotation. Such a lever would not work well as a chemical balance because it would require that equal weights be placed simultaneously on each pan of the balance—which is impossible since initially one of the weights is unknown! Thus, a chemical balance beam is constructed so that most of its mass lies below the level of the knife edge that is supporting it, and its center of gravity lies at a significant distance directly below the point of rotation. If a small unknown weight is placed on, say, the right-hand balance pan, the beam will start to rotate clockwise, but in doing so it will have to lift to the left the center of gravity of the beam itself, which creates an opposing counterclockwise torque. This opposing torque becomes greater the farther the beam is tilted (while the torque of the unknown weight becomes smaller), so that, at a modest tilt, the beam can come to a new equilibrium just a little removed from the one it had before the unknown weight was placed on the pan. If perturbed by a small force from outside, the chemical balance responds in such a way as to relieve or oppose the force that is disturbing it.

A chemical equilibrium is itself like a chemical balance. In a typical chemical reaction, where reactant molecules A and B come together to produce end products C and D, the end products always have a tendency (even if sometimes very small) to reverse the reaction and so revert to the original reactants. The forward reaction has a rate, v_1, that is proportional to the concentrations of A and of B, while the reverse reaction has a rate, v_2, that is proportional to the concentrations of C and D, the respective concentrations being denoted by [A], [B], [C], and [D]:

$$A + B \rightarrow C + D \qquad v_1 = k_1 [A][B]$$
$$A + B \leftarrow C + D \qquad v_2 = k_2 [C][D]$$

where k_1 and k_2 are "rate constants" for the forward and reverse reactions. At equilibrium, the rates of the forward and reverse reactions are equal:

$$k_1 [A][B] = k_2 [C][D]$$

This leads to the important fact that equilibrium requires (at a given temperature and pressure) a specific ratio of reactants to end products:

$$\frac{k_1}{k_2} = \frac{[C][D]}{[A][B]} = K_{eq}$$

where K_{eq} is called the equilibrium constant for the reaction.

The equilibrium is dynamic: the constant ratio of end products to reactants is maintained because $v_1 = v_2$, not because v_1 and v_2 go to zero. Individual molecules are converted back and forth across the reaction boundary, but there is no net change. This dynamic nature of the equilibrium was first directly demonstrated by George Hevesy (1913), working in the laboratory of Ernest

Rutherford in Manchester. Hevesy had been asked by Rutherford to separate what was then called radium D (a radioactive product of the natural breakdown of radium, or ultimately of uranium-238) from all the nonradioactive lead-206 in which it was embedded. Hevesy was a skilled analytical chemist, and when all his attempts to separate radium D from lead failed, he decided that the two were chemically indistinguishable, that radium D is a radioactive isotope of lead (lead-210, as it turned out). It occurred to him that he could use his radioactive lead to label ordinary lead and to conduct experiments on lead that could not previously be done. These were the first experiments ever performed using radioactive tracers. In one experiment, Hevesy placed crystals of radioactive lead chloride and nonradioactive lead nitrate in contact with a common solution saturated with ordinary lead, nitrate, and chloride. He found later that the crystals of lead nitrate had become radioactive, thus showing that there had been interchange between the lead of each of the crystals and that of the solution. The dynamic nature of a multitude of chemical and biochemical equilibria has subsequently been demonstrated by isotopic experiments.

Because a chemical equilibrium is dynamic, it represents a balancing act among all the chemical components involved. If, to an equilibrium involving A, B, C, and D, a little extra D is added, the system responds by reacting some of the D with C to form A and B—just enough to restore the ratio of products to reactants back to the value of the equilibrium constant. Likewise, the addition or subtraction of any of the components to the system will invoke a response such that the system reduces the change just made to it. If the reaction $A + B \rightarrow C + D$ involves the production of heat (thermal energy), then heat can be regarded as an end product of the reaction, and the addition of heat from outside by raising the temperature will provoke the equilibrium to reverse the reaction a little to use up some of the heat (and the equilibrium constant itself will be somewhat reduced). If a reaction involves the production of a gas, thereby increasing the system's volume, an increase in ambient pressure will cause the system to reverse a little to relieve the pressure.

The effect of temperature on the equilibrium constant was formulated by Jacobus van't Hoff in the 1880s, but van't Hoff's results were publicized by Henri Le Chatelier in 1884, and the idea that a chemical equilibrium responds to temperature, pressure, and chemical concentrations in such a way as to reduce the provocation has come to be known as *Le Chatelier's principle*. The principle gives us the direction but not the size of the response of an equilibrium system. We will apply it to two examples: the effects of temperature and pressure on an equilibrium between ice and water, and the effect of temperature on a stretched rubber band.

We previously discussed the ice–water equilibrium as an example of the phase rule—how the system can have only one degree of freedom, so that if the pressure is changed, the temperature has to change as well in order to maintain a balance between the free energies of ice and water. Pressure and temperature are not independent variables if an equilibrium is to be maintained, but must vary together (in opposite directions). What does Le Chatelier's principle say about the ice–water system? Its prediction about temperature is no surprise. If temperature is increased, ice melts because, in doing so, it absorbs entropy (and with it thermal energy), thus tending to lower the temperature. If temperature is lowered, water freezes because, in doing so, it liberates entropy (and thermal energy), thus tending to raise the temperature. This works so long as the temperature change is not too prolonged, so that ice is not completely melted nor the water completely frozen. With regard to the effect of pressure, the Le Chatelier principle predicts that because a gram or mole of ice has a greater volume than an equal mass of water, increased pressure causes ice to melt, which tends to relieve the pressure by reducing the volume of the system. A reduction of pressure causes water to freeze, which tends to restore the pressure by increasing the volume of the system. As with temperature changes, the pressure changes cannot be too prolonged or the system may completely melt or completely freeze. The phase rule dictates that temperature and pressure must change together in a specific way if equilibrium between ice and water is to be maintained.

What is the specific quantitative way that temperature and pressure changes must be related to one another? Émile Clapeyron (1834) derived a relation for water and steam, but his expression, known today as the *Clapeyron equation*, can also be applied to the water and ice system or to any two phases in equilibrium with one another. For equilibrium, molecules of H_2O must have the same tendency to escape from ice to water as they have to escape from water to ice. If the escaping tendencies of ice and water molecules are the same, their chemical free energies are said to be the same, and there is consequently no free energy change in passing from one phase to the other. We will see later that the chemical free energy, G, of a substance increases with pressure, P, in proportion to its volume, V, and decreases with temperature, T, in proportion to its entropy, S:

$$\frac{dG}{dP} = V$$

$$\frac{dG}{dT} = -S$$

The ice has more volume (per gram or mole) than the water, while the water has more entropy than the ice. When pressure increases, free energy of the ice *increases* more than that of the water, causing the ice to melt. When temperature increases, the free energy of the water *decreases* more than that of the ice, again causing the ice to melt. So, to keep the free energies of ice and water equal to each other, an increase in pressure has to be accompanied by a decrease in temperature. Let ΔV be the volume difference between ice and water (a positive number), let ΔS be the entropy difference between ice and water (a negative number), and let ΔP and ΔT be the changes in pressure and temperature to which the ice and water system are subjected. Then:

$$\Delta P \cdot \Delta V = \Delta T \cdot \Delta S$$

Since ΔV is positive (for this case) while ΔS is negative, if ΔP is positive then ΔT must be negative so that both the left-hand and right-hand sides are positive. The ratio of pressure change to temperature change is:

$$\frac{\Delta P}{\Delta T} = \frac{\Delta S}{\Delta V}$$

If the right-hand side is multiplied top and bottom by T, we have $T\Delta S$ in the numerator, which is the "enthalpy" (thermal energy), ΔH, which must be added to ice to convert it to liquid water:

$$\frac{\Delta P}{\Delta T} = \frac{\Delta H}{T\Delta V}$$

From the value of ΔH, the thermal energy required to melt ice (about 80 calories per gram), and the value of ΔV, the difference in volumes of equal masses of ice and water, the ratio of changes in pressure and temperature (at a given temperature) can be determined.

For the ice–water equilibrium, it is almost as if (to use a fanciful aid to memory) the ice has too much volume, which it would like to get rid of to turn into water, while the water has too much entropy, which it would like to get rid of to turn into ice. If the ice is helped by an increase in pressure to get rid of its volume, then the water must be helped by a decrease in temperature to get rid of its entropy, and, if both occur in the right ratio to one another, neither ice nor water wins, and both remain in equilibrium with one another.

That a single substance such as water can exist in two (or indeed three) forms side by side at equilibrium with one another, sharing the same environment of temperature and pressure, is a remarkable phenomenon. It is not easy to imagine a world without this great buffering capacity that transforms what otherwise would be explosive events into gradual changes. Life obviously would be impossible if our blood, subject to a momentary drop in atmospheric temperature or pressure, were suddenly to turn to ice or, subject to a slight fever, were to vaporize at once into the air.

A cube of ice next to a cube of water would seem to constitute a system exhibiting a striking discontinuity and heterogeneity. We have noted that change arises from differences, yet here is a difference that will never give rise to a change. The ice and water are content to repose forever together without causing anything to happen, so long as temperature and pressure are right. A temperature of 0°C at atmospheric pressure is just fine for preserving them side by side, even though they have such different properties. In energetic terms, their two striking differences are that the ice has more volume than the water (gram for gram, or mole for mole), and the water has more entropy than the ice. Because the water has more entropy than the ice at the same temperature, it has more thermal energy than the ice, but this energy is bound within the constraints of the system; that is, it can do no work.

Because a molecule of H_2O can as easily pass from ice to liquid as from liquid to ice, the two phases are said to have the same free energy, which is to say that there is no free energy change involved in moving H_2O from one phase to the other.

Note, however, that the bound energy of the water (or ice) can become free energy in a different context, and the fact that water and ice have the same free energy in relation to one another does not mean that they have the same free energy in relation to something else.[2] Suppose that a gram of water at $0°C$ (A) and a gram of ice at $0°C$ (B), are each placed (separately) in contact with a gram of ice at $-10°C$ (C). The free energy difference between A and C is greater than the free energy difference between B and C, even though the free energies of A and B, in relation to one another, are the same. By virtue of its higher entropy content, A can contribute more entropy (and thermal energy) to C before equilibrium is reached than can B, and the respective equilibria will be different. The ice–ice system will reach an equilibrium of pure ice at a temperature of about $-5°C$, but the water–ice system will reach an equilibrium that is about 7/16 water and 9/16 ice, all at $0°C$. About 10 calories of thermal energy will pass from A to C across a somewhat higher average gradient than the approximately 5 calories that pass from B to C. A clever engineer of the future might figure out how to get more work out of the water–ice system than the ice–ice system.

Another example of the application of Le Chatelier's principle is seen in the behavior of a rubber band. If a relaxed (unstretched) rubber band is heated, it will expand, as most substances do. But

---

2. This statement may sound strange to someone used to the Gibbs (or Helmholtz) chemical free energy. Free energy as used here applies to any system containing a potential (gradient), including one that has a thermal gradient. The Gibbs (or Helmholtz) free energy, which is used in chemistry (and which we will use in Chapter 7), applies to isothermal chemical systems containing no thermal gradient.

if the rubber band is first stretched and then heated, it will contract. This seems very peculiar. When I was a student of physical chemistry, our professor, Arthur Campbell, demonstrated an upright wheel on a horizontal axis that had a lightweight rim strung with stretched rubber bands as spokes. When a heat lamp was shined on the wheel asymmetrically, so that the light hit the rubber spokes on one side of the hub but not the other, the wheel began to rotate, and continued to do so as long as the light was on. The stretched rubber bands that came under the light contracted enough that the torque of gravity on the unheated side became greater than on the heated side, pulling one side of the wheel down more than the other and causing it to rotate. If it had been focused sunlight rather than an electric lamp, it would have been a solar-driven wheel. But why does stretched rubber contract when heated? Le Chatelier's principle provides one kind of answer. When rubber is stretched, work is done in pulling it out, and the rubber warms slightly, somewhat as air warms in a tire pump when work of compression is done upon it. If the rubber can contract against an opposing force, it will do work and will cool. So, if the stretched rubber is heated, its contraction can be viewed, via Le Chatelier's principle, as an attempt to cool itself and thus relieve the provocation of the heat. This is not the sort of answer that will satisfy someone interested in the molecular structure of rubber, but it provides one kind of insight on a rather peculiar phenomenon.

Chapter 5

Energy and Entropy in Heat Engines and Heat Pumps

THE MACHINE AS METAPHOR FOR NATURE

Our modern age is well aware of the importance of science to the development of technology and industry, but we are prone to forget that science has also been aided by practical affairs, especially in earlier times. As new mechanical devices were invented and became more common in medieval and early modern times, they gave people new ways of thinking about how nature works. It is unlikely that William Harvey would have conceived of the heart as a pump if he had not seen many pumps in use in London, including the waterworks at London Bridge and near "Broken wharfe," which supplied water to the region where he worked. From his familiarity with pumps, he was able to imagine that the right ventricle, with its tricuspid and pulmonary valves, was pushing blood through the lungs "as by two clacks (valves) of a water bellows to rayse water." He conceived the rapid sequence of events at the heart—the movements of auricles, then ventricles, then arteries—as being like the action of one gear driving a sequence of others in a watermill, or to the complex firing mechanism of a flintlock gun.

Medieval Europe saw increasing use of water and wind power to relieve human and animal labor, especially in pumping water and grinding grain. The way was frequently led by the monasteries and abbeys, which were often built on a stream, which supplied the power of running water as well as the water itself. The Domesday Book of William the Conqueror reported 5,624 water mills in England, most of them undershot wheels driven directly by the current of a stream. Here was an early lesson on the importance of gradients, for such a wheel obviously would not turn in still water, and its power depended upon the speed of the water flow, which in turn relied on the drop in elevation of the stream. If an abbey downstream built a dam to drive a new overshot wheel, the monks upstream would be upset, for such an elevation of the water downstream reduced their gradient. The power of their waterwheel depended upon both the upstream and downstream elevations of the water, upon the difference in elevation across which their water was running.

By the end of the 17th century, mathematically astute engineers were making use of the new physics of Galileo, Newton, Leibniz, and Huygens to think about how fast a water wheel should turn, and against what external load it should work, in order to capture the maximum *vis viva* of the running stream. As a pendulum gained enough *vis viva* on its downward swing to carry itself back to its original height, so water dropping across a height should ideally be able to lift its own weight to that height. The efficiency of a water wheel in lifting a weight could be measured in such terms. In 1704, Antoine Parent published a mathematical analysis suggesting that the efficiency of an undershot waterwheel could not exceed 14.8%. John Smeaton in 1759 thought otherwise and estimated that an undershot waterwheel could reach efficiencies of 33–50%, and an overshot wheel might attain 66–100%. But, by Smeaton's day, new machines were being built in England, powered by fire rather than running water.

GETTING WORK FROM FIRE

How do we get fire to do useful work? Denis Papin in 1691 had the key idea. He showed that the steam generated from water placed at the bottom of a cylinder could push a piston up when the water was heated by fire, and, when the fire was removed and the steam condensed, the atmosphere would push the piston back down. Thomas Newcomen built the first successful steam engine based on the idea of first expanding and then condensing steam under a piston. Perhaps as early as 1705, the first Newcomen engine was used in Cornwall for driving a pump to raise water from a flooded mine, and, in 1712 an improved version was operating in Birmingham. In the Newcomen engine, an oscillating beam attached to the piston was counterweighted so that gravity pulled the piston up and drew steam from a boiler into the cylinder. When the cylinder was sprayed with cold water, the steam condensed and atmospheric pressure pushed the piston down. It was the atmospheric pressure, rather than the steam, that created the power stroke; the steam served for driving the air from the cylinder so that, upon condensation, a partial vacuum was created.

James Watt worked as an instrument maker at Glasgow University, where he was a friend of the chemist Joseph Black, who formulated the calorimetric means for defining quantities of heat and who, for phase changes, conceived the idea of latent heat. Watt was asked to determine why a scale model of a Newcomen engine was not working properly. In his investigations, Watt invented many improvements to the Newcomen engine and eventually went into business with Matthew Boulton manufacturing the new Watt-Boulton engine, which became an important part of the Industrial Revolution in England. One key improvement was the use of a cooling, condensing cylinder separate from the piston-driving cylinder, so that steam could be condensed without heat losses associated with cooling the main cylinder. Another improvement was a means for cutting off the steam from the boiler before the piston was fully driven forward, so that steam already in the cylinder could further expand without added heat and thus begin to cool. This divided the power stroke into two phases, the first with heat and the second without, phases that would later be idealized as "isothermal expansion" and "adiabatic expansion" of the steam.

The invention of the steam engine and the initiation of the Industrial Revolution occurred in England, which was rich in coal but rather poor in science during the second half of the 18th century. Both the Royal Society and the English universities had fallen far behind their counterparts on the continent. Yet when the French, with their distinguished École Polytechnique and brilliant array of mathematicians, physicists, and engineers, awakened from the Napoleonic Wars, they found an England that had far outrun them in industrialization and in the invention of fire-driven machines. One young French engineer, determined to understand the physical principles of the steam engine, was Sadi Carnot (1796–1832), who, in 1824, published a small book entitled *Reflections on the Motive Power of Fire*, a book which the eminent historian of technology, Donald Cardwell, has called "the most *original* work of genius in the whole history of the physical sciences and technology."

SADI CARNOT: TEMPERATURE GRADIENTS AND REVERSIBLE CYCLES

Sadi Carnot was the son of Lazare Carnot, a distinguished French engineer who employed the new mathematics and physics in the analysis of machines. An efficient machine—one that delivers the most effect from the power received—is a smooth and quiet machine, with no leakage of power, with minimized friction, and without wasted motion or impacts of one part upon another. The coupling of power source to ultimate action is so finely balanced as to suggest—at least if your name is Carnot—that the ideal machine might be run in reverse by a slight change in forces. Whether he invented the idea himself or got it from his father, Sadi Carnot regarded reversibility as a necessary condition for a machine of the highest possible efficiency.

Carnot explicitly compared the steam engine to an overshot waterwheel, with caloric (heat) falling across a temperature difference in a steam engine being like the water falling across a height difference in the waterwheel:

> We are sufficiently justified in comparing the motive power of heat with that of a fall of water. The motive power in both cases has a maximum value which cannot be exceeded, no matter what engine is used to harness the action of the water in the one case and no matter what substance is used to harness the action of the heat in the other. The motive power of a fall of water depends on its height and on the amount of liquid. The motive power of heat likewise depends on the amount of caloric that is used and on what might be termed— in fact on what we shall call—the height of its fall; it depends, in other words, on the difference in temperature of the bodies between which the passage of caloric occurs.[1]

In this passage, Carnot has pointed out that more than a fire is necessary to drive a steam engine—that the engine must operate across a difference of temperature and therefore must have not only hotness, but coldness as well. The Watt (or later Woolf)

1. Sadi Carnot, *Reflexions on the Motive Power of Fire*. Translated and edited by Robert Fox. Manchester: Manchester University Press, 1986. p.72.

engine with its cold, condensing cylinder as well as its hot, power cylinder, may have helped to drive home this reality, which is not an obvious point to an untutored observer.

Fire derives its motive power from its ability to make things expand, but expansion alone is not enough to sustain a power output. Expansion must be followed by contraction, and then expansion again, and contraction again. The engine must run in cycles of expansion and contraction, and while the expansion requires hotness, the contraction requires coldness. Hence, two different temperatures are required to allow the engine to run in cycles of repetitive action, and the greater the difference in temperature, the more forceful and rapid the action can be. Carnot had the genius to see that steam was not really required for a heat engine—that air or any other gas could provide the cycles of expansion and contraction on heating and cooling. And he saw that, in running in cycles of expansion and contraction, the gas was continually returning to a state that it had previously occupied.

Carnot saw that in the Watt (or Woolf) engine, the expansion and contraction phases of the cycle are each divided into two parts, so that there are four parts in one complete cycle:

1. Expansion of the gas while heat flows into it from a hot source. The gas is considered to be at the temperature of the hot source and to maintain that temperature throughout this "isothermal expansion" phase. The gas does work on the piston, but does not cool because it receives heat from the hot source.

2. Further expansion of the gas after the hot source has been cut off. The gas is considered insulated from thermal exchange, so that this is an "adiabatic expansion." The gas cools as it does further work on the piston.

3. Compression of the gas as the piston reverses and does work on the gas, but the gas remains at the same temperature as in Step 2 because the heat generated by compression is drained into the cold source in this "isothermal compression."

4. Further compression of the gas after the cold source has been cut off, so that in this "adiabatic compression" the temperature of the gas is raised. At the end of this Step 4, the gas and the piston are in the same state as at the beginning of Step 1.

Expansion of a gas can produce work as it pushes on the piston, while compression of the gas requires work as the piston pushes on the gas. Steps 1 and 2 in the Carnot cycle therefore can do work, while Steps 3 and 4 require work, and the net work accomplished in going once around the cycle is given by adding the work done in Steps 1 + 2 and subtracting that required in Steps 3 + 4. But is there any net work in fact accomplished? The adiabatic Steps 2 and 4 are operating between the same two temperatures (hot and cold) and are the reverse of one another, so that, ideally, whatever work is produced in the adiabatic expansion of 2 is used up in the adiabatic compression of 4. The key to a net production of work lies in the difference between the two isothermal steps. The isothermal expansion Step 1 is done at high temperature, while the isothermal compression Step 3 is done at low temperature. The high temperature expansion can do more work than is required by the low temperature compression, and thus the overall cycle can produce work.

Carnot's book was published two decades before the law of the conservation of energy was formulated, but there is a conservation principle implicit in his concept of reversibility that may have contributed to the thought of Hermann Helmholtz.

Carnot thought, in 1824, that a heat engine utilized falling caloric to produce work, just as a waterwheel made use of falling water. Caloric, he thought, was conserved (after publication, he altered this view). Reversibility implies that whatever work is produced from falling caloric must be just sufficient to pump that same caloric back up to the temperature from which it started. A reversed heat engine would become a caloric pump, and the machine might in theory go back and forth, doing one or the other without loss. In reversibility, Carnot had found an optimum of performance beyond which no heat engine could go, for to have an engine of greater performance paired with this ideal would be to claim the ability to create continual power from nothing, or to pump caloric from low to high temperature without the expenditure of work.

Carnot also surmised that the theoretical limit to the efficiency of a reversible heat engine was set by the temperature difference across which it operated and not by the particular gas (steam, air, etc.) that was used in the chamber. This was a remarkable conclusion. The importance of the thermal gradient had been discovered, together with the ideal of reversible operation in which work and flow of caloric were convertible one into the other, either way, with conservation of their sum implied.

Èmile Clapeyron (1834) developed Carnot's ideas mathematically and depicted graphically the Carnot four-part, expansion-compression cycle. Carnot died of cholera at age 36, 2 years before Clapeyron's memoir, and Carnot's own book became hard to find. While studying in Paris in 1845, William Thomson (later Lord Kelvin), then only 21 years old, came across Clapeyron's paper, and through it learned of Carnot's book, but he could find no Paris bookseller who had heard of Carnot. Both William Thomson and his German contemporary Rudolf Clausius initially knew of Carnot's work only through Clapeyron's writings.

According to Carnot (and Clapeyron), the steam engine produced work by utilizing the motive power of caloric falling across a temperature gradient. If caloric could be translated as "entropy," we would have the modern view, but, in Carnot's day, entropy was not a word, and caloric was taken to mean that which could be measured calorimetrically as calories—in other words, it was what the English called "heat," which was about to be shown by Joule to have an equivalence to mechanical work. Carnot portrayed caloric as conserved in the steam engine, just as water is conserved in a waterwheel. The work of the engine was produced from falling caloric, not from caloric itself (though, after publication, Carnot began to doubt the conservation of caloric).

RUDOLF CLAUSIUS AND THE FORMULATION OF ENTROPY

With the acceptance of the conservation of energy, which said that the sum of mechanical work and heat was constant, there arose a problem: if heat was conserved in the steam engine, yet work was produced, the steam engine appeared to violate the new law of energy conservation. Rudolf Clausius (1822–88) took up the challenge of resolving this contradiction between Carnot and energy conservation.

Clausius concluded that heat cannot be a conserved quantity for it is produced from purely mechanical action, as in Rumford's boring of cannon or in just rubbing one's hands together. Accepting Joule's demonstrations that the sum of work and heat is conserved, Clausius argued that some of the heat driving a steam engine is converted to work. Clausius regarded heat as a kind of motion of microparticles that causes expansion, but, in a sense, he did not entirely free himself of the old view that it is a kind of subtle fluid—for heat remained the unitary stuff that flowed across the temperature difference of the steam engine.

Because he regarded heat as flowing across the temperature difference, he wondered how much work can be obtained from a unit of heat flowing down a unit of temperature difference. He concluded that there is no constant value—that a unit of heat is converted to work less effectively when it passes across a unit of temperature difference at high temperatures than at low (see our previous discussion of this point in Chapter 3). Thus, he decided that the "equivalence value" of the heat (the ratio of work produced divided by heat) is less at high temperature than at low. He therefore divided heat by a function of temperature (later the absolute temperature itself) to get the true equivalence value of heat at all temperatures. When multiplied by the temperature difference, the equivalence value represents the maximum work that can be obtained from the heat:

$$w = \left(\frac{q}{T_1}\right)(T_1 - T_2) = q\left(\frac{T_1 - T_2}{T_1}\right)$$

where $\dfrac{q}{T_1}$ is the equivalence value of the heat at temperature T_1, and w is the maximum work that can be obtained from heat q.

But if, according to Clausius, a quantity of heat moves from one temperature to another, then a transformation has occurred because the equivalence value of the heat has changed, and Clausius speaks of "the equivalence value of the transformation," which is the difference in the two equivalence values of the heat: $\dfrac{q}{T_2} - \dfrac{q}{T_1} = q\left(\dfrac{1}{T_2} - \dfrac{1}{T_1}\right)$. Clausius then starts to use the term "transformation" for this difference, and he notes that, in a cyclic process, the total transformation, being the sum or integral of these differences, must be positive, unless the process is reversible in Carnot's sense, and thus completely compensated, without any heat being wasted. Clausius then introduced N to represent the integral of transformations but shortly changed its sign to $-N$. Finally, he introduced S to stand for the equivalence value $\frac{q}{T}$, and named it "entropy," for "the transformation within."

In summary, Clausius preserved "Carnot's principle" that work is produced in a steam engine by heat falling across a temperature difference, but he reconciled this principle with the conservation of energy by saying that an amount of heat disappears equal to the amount of work produced. Instead of heat being conserved, entropy is conserved—but only in the ideal reversible engine. In any real engine in which part of the work is wasted by friction or part of the heat is wasted by leakage, entropy increases.

Unfortunately, Clausius's brilliant analysis left (and still leaves) many of his readers numb with incomprehension and confusion. Clausius had discovered the compound nature of thermal free energy before anyone was ready to see it in that light. Entropy entered the world of science not as the extensive factor in thermal energy, but as a mathematical construct formulated as $\frac{q}{T}$, a quantity that needed always to increase, for reasons that few outside the inner circle could fully understand. His statement (1865) of the second law, "The entropy of the universe tends to a maximum," is rightfully repeated in every textbook, but the meaning of entropy has confounded generations of students because of the peculiar historical way it came into being.

THE IDEAL EFFICIENCY OF HEAT ENGINES AND HEAT PUMPS

The efficiency of a heat engine is defined as the ratio of the work produced to the heat consumed. The principle of the conservation of energy implies that work and heat are measurable in the same unit, a common unit of energy such as the joule or the calorie, and that therefore the efficiency ratio is a dimensionless quantity.

In a heat engine, the thermal free energy is the product of its temperature difference and its entropy. The free energy *is* the maximum work that can be produced under ideal (reversible) conditions—conditions in which the decrease in thermal free energy is completely compensated by the increase in some other free energy, such as mechanical work. For a unit of entropy, S, the maximum work, w, which can be produced is:

$$w = S(T_1 - T_2)$$

where T_1 and T_2 are the high and low temperatures, respectively. For that same unit of entropy, the thermal energy, q, taken from the hot reservoir is:

$$q = ST_1$$

The ratio of maximum work to thermal energy is therefore:

$$\frac{w}{q} = \frac{S(T_1 - T_2)}{ST_1} = \frac{T_1 - T_2}{T_1}$$

This is the maximum efficiency possible for an ideal, reversible heat engine. If, for example, a reversible heat engine operated between the boiling and freezing points of water, its efficiency would be $\dfrac{373 - 273}{373} = 0.268$ or 26.8%.

With the reversible engine, the entropy of the system remains constant, as a given amount of entropy simply falls from high temperature to low, all the time reversibly coupled to the increase of some other (work) gradient. But if the engine is not reversible, if all the thermal free energy is not converted to work, the entropy will increase. Suppose, for one case, that some of the mechanical work that might have been produced is lost in the friction of the piston in the cylinder. The friction produces heat (thermal energy), which is entropy times the temperature at which it is produced, that is, TS. This is new entropy, additional to that which flowed across the $(T_1 - T_2)$ gradient, so that the entropy of the system has increased. Or, for a second case, suppose that some entropy flow (and accompanying heat) is short-circuited from the hot reservoir to the cold without contributing to the expansion of gas in the cylinder, so that this extra entropy flow does no work. The entropy flow itself remains within the system, so that, in itself, it does not seem to contribute to an entropy increase. But an irreversible flow of entropy creates entropy (as we have seen before), just as does an irreversible flow of electrical charge. If a unit of extra heat q_e is short-circuited between T_1 and T_2, it requires less entropy to accompany this heat at the higher T_1 than at the lower T_2: that is, q_e is $T_1 S_1$ as it leaves the hot reservoir, but $T_2 S_2$ as it enters the cold reservoir:

$$q_e = T_1 S_1 = T_2 S_2$$

$$S_2 = S_1 \left(\frac{T_1}{T_2} \right)$$

Since T_1 is greater than T_2, S_2 is greater than S_1, which means that entropy has increased during the irreversible conduction of q_e from the high temperature to the low.

Note that the efficiency of an ideal, reversible heat engine goes to zero if T_1 equals T_2 because there is no free energy when there is no gradient, and hence no work can be produced whatever the temperature. Work cannot be produced from an equipotential thermal system. It is no use trying to get work out of a warm bath—unless you have a cold bath beside it.

A warm-blooded animal produces lots of thermal energy, but its temperature is nearly uniform, at least in its central core. The internal environment of each of its billions of cells has almost the same temperature as that of the cells. Clearly, the cells are not acting as heat engines in the way that they process free energy, else their efficiency would be essentially zero.

A heat pump is (as Carnot foresaw) a heat engine in reverse. Instead of using a temperature difference to create mechanical or electrical work, it uses mechanical or electrical work to create a temperature difference. Performance is assessed in this case not as the ratio of work produced to heat transported, but as the ratio of heat transported to work required. The work performed by the pump is that of pushing entropy up a temperature gradient, $w = S(T_1 - T_2)$, while the heat transported is the same entropy deposited at the higher temperature, giving $q_h = ST_1$. Then the "coefficient of performance" for an ideal heat pump, representing the maximum units of thermal energy that can be delivered (to a hot region) by a unit of mechanical (or electrical) work, is:

$$\frac{q_h}{w} = \frac{ST_1}{S(T_1 - T_2)} = \frac{T_1}{T_1 - T_2}$$

The temperature difference, $T_1 - T_2$, is the friend of the heat engine but the enemy of the heat pump, which is why heat pumps for heating houses are more economical and popular in mild climates than in cold ones, and also why it is highly beneficial in cold winter climates to have a heat exchanger that remains above freezing at the bottom of a pond or deep underground instead of having to work against air temperatures that go well below freezing. For a house kept warm by a heat pump operating between temperatures of 35°C (at a heater inside the house) and 0°C outside, the coefficient of performance for an ideal pump would be $\dfrac{308}{308-273} = 8.8$. Although a real heat pump would not achieve this ideal limit, it might deliver several times more heat, from the same electrical energy, as a conventional electric heater inside the house.

A refrigerator or air conditioner is a heat pump, but one in which the objective is to take heat from a cold region rather than to deliver it to a warm one. The coefficient of performance is assessed accordingly, with the low temperature, T_2, replacing the high temperature, T_1, in the numerator of the previous expression for the coefficient of performance:

$$\frac{q_c}{w} = \frac{ST_2}{S(T_1 - T_2)} = \frac{T_2}{T_1 - T_2}$$

The ratio represents the maximum units of thermal energy that can be removed (from a cool region) by a unit of mechanical (or electrical) work.

The ideal coefficient of performance for a refrigerator or air conditioner is less by 1 than the ideal coefficient of performance for the heat pump operating between the same temperatures:

$$\left(\frac{T_1}{T_1 - T_2}\right) - \left(\frac{T_2}{T_1 - T_2}\right) = \frac{T_1 - T_2}{T_1 - T_2} = 1$$

The efficiency of the reversible heat engine can also be derived from the behavior of an ideal gas undergoing the four parts of the Carnot cycle. We have seen that the free energy change, and hence the maximum work that can be obtained, from the isothermal expansion of an ideal gas is $w = RT \ln \frac{V_2}{V_1}$. This is also the work that must be done on the expanded gas to compress it to its original volume. The isothermal expansion of Step 1 yields a maximum work of:

$$w = RT_1 \ln \frac{V_2}{V_1}$$

which is also the amount of heat, q_1, which must be conducted from the hot reservoir in order to keep the expansion isothermal.

The isothermal compression of Step 3 requires work of:

$$w' = RT_2 \frac{V_3}{V_4}$$

which is also the amount of heat, q_2, which must be conducted to the cold reservoir in order to keep the compression isothermal.

The net work from these two isothermal steps is $w - w'$. It will be shown later that $\frac{V_3}{V_4} = \frac{V_2}{V_1}$, so that the net work from the iso-thermal steps is $R(T_1 - T_2) \ln \frac{V_2}{V_1}$.

Since the adiabatic expansion of Step 2 and compression of Step 4 operate across the same temperature difference, they cancel each other out because the work produced in the first equals that required in the second. The efficiency of the entire (reversible) cycle for an ideal gas is:

$$\frac{w}{q_1} = \frac{R(T_1 - T_2)\ln\dfrac{V_2}{V_1}}{RT_1 \ln\dfrac{V_2}{V_1}} = \frac{T_1 - T_2}{T_1}$$

Thus, the efficiency calculated from the behavior of an ideal gas in the Carnot cycle agrees with that calculated from entropy flow across the thermal free energy gradient.

To show that the preceding volume ratios are indeed equal to one another, consider the work done in an adiabatic expansion:

$$PdV = \frac{RTdV}{V}$$

This work causes the internal energy of the gas to decrease as its temperature drops from T_1 to T_2 during the expansion. The energy decrease is also given by the heat capacity of the gas times the change in temperature: $C_v dT$, as this is the amount of thermal energy that would restore the temperature and internal energy of the gas. The two energy terms, for work and for heat, are equal:

$$C_v dT = \frac{RTdV}{V} \quad \text{or} \quad \frac{C_v dT}{T} = \frac{RdV}{V}$$

$$C_v \int_{T_2}^{T_1} \frac{dT}{T} = R\int_{V_1}^{V_2} \frac{dV}{V}$$

$$C_v \ln\frac{T_1}{T_2} = R\ln\frac{V_2}{V_1}$$

Since two adiabatic expansions or compressions sharing the same temperature differences experience the same internal energy change as expressed on the left-hand side of this equation, they must also share the same work terms as expressed on the right-hand side, and therefore their volumetric ratios must be the same.

Chapter 6

The Spreading Tendency of Nature

BROWNIAN MOTION AND DIFFUSION

The concept of energy emerged, as we have seen, in the 1840s with the realization by Mayer, Joule, Helmholtz, and others of the intimate equivalence of mechanical work and heat. Since the time of Liebniz, it had been speculated that when mechanical motion disappears in friction or in collisions, the motion of the large object that we can see is converted to the microscopic motion of fine particles we cannot see. If mechanical motion becomes heat, as Joule demonstrated in his quantitative experiments, overall motion (or energy) remains conserved because heat is, in the words of Rumford, "a kind of motion."

It seems paradoxical that the principle of the conservation of energy grew partially out of the human experience that machines are incapable of perpetual motion, that they cannot keep running by their own devices alone—yet for energy to be conserved when macroscopic motion is converted to the microscopic motion of heat, it is necessary to suppose that the motion of heat goes on forever. In other words, the perpetual motion that is forbidden in the world that we directly experience is required in the world that lies beneath the fabric of our senses.

THE DISCOVERY OF BROWNIAN MOTION

In the summer of 1827, about 15 years before Robert Mayer's first paper on energy conservation, this microscopic perpetual motion was seen by Robert Brown, looking through a high-magnification, newly developed, simple achromatic lens at the British Museum. Brown made many microscopic discoveries in botany. It was he who first described and named the "nucleus" of living cells and first observed the remarkable cyclic motion of green chloroplasts within certain plant cells—a phenomenon known as *cytoplasmic streaming* or *cyclosis*. He also distinguished reproductive differences between gymnosperms (the conifers and cycads) and angiosperms (the true flowering plants). But what he saw that summer in 1827 was more amazing and mysterious than any of his other discoveries.

Brown was studying the reproductive process in specimens of *Clarkia pulchella*—an exotic wildflower from the American northwest that had first been collected by Meriwether Lewis of the Lewis and Clark expedition and named for his co-leader, William Clark. Brown saw that small particles inside the pollen grains from *Clarkia pulchella* were in constant motion, jiggling from side to side and sometimes rotating. Each particle had its own motion, independent of its neighbors, without any concerted movement of particles in one direction, and could not be explained by currents caused by heat or evaporation or vibrations of the microscope itself. Others gazing through microscopes had noticed motion in very fine particles, but Brown was the first to investigate it in detail, and, since his time, it has been known as "Brownian motion."

You may have seen dust particles drifting erratically in the air in a beam of sunlight, darting up and down, right and left. Brownian motion is a little reminiscent of that, yet very different.

For the dust particles move in groups, blown by minute currents of air, and eventually they settle on someone's bookshelf or windowsill, while the Brownian particles shimmer individually, and their motion goes on forever. As Jean Perrin remarked in his essay "Brownian Motion and Molecular Reality" (1909), "the most striking feature of Brownian movement is the absolute independence of the displacements" and, he added, "what is really strange and *new* in the Brownian movement is, precisely, that it never stops."

Anyone with a beating heart will, on first seeing this extraordinary microscopic sight, be excused for feeling for a moment that he or she is looking into the soul of nature—though I have also seen students whose demeanor seemed to say "so what?" A natural reaction is "The particles must be alive," a feeling that emerges from some primal instinct, common to the animal world, that unexpected motion must be a sign of life, a fundamental "*vis viva.*" Brown at first thought he was witnessing a living phenomenon— after all, he was looking at particles within a living plant. Because the particles were inside pollen grains, he supposed at first that they might be involved in the fertilization of the ovum. But they continued to move even when released from the pollen into water. The phenomenon was not specific to the exotic American plant, for the pollen from other flowers also had moving particles—as did, extending the search, spores from moss and *Equisetum* (a fern ally), plants which in his time were believed (incorrectly) not to produce sexual gametes. The motion had nothing to do with sex! Even the vegetative parts of plants could produce moving particles. Nor need the plants even be alive, as material preserved in alcohol, or herbarium specimens dried a hundred years before, or fossilized wood could generate moving particles. If this was a living phenomenon, life had a long memory.

Charmed by the mystery of his extraordinary findings, Brown extended his study to the realm of minerals and all sorts of non-living materials. The London soot vibrated under his microscope, as did tiny bits of broken window glass and stony grains from a piece of the Sphinx (remember he was working at the British Museum). Asbestos, manganese, lead, granite, and many other substances were ground up to produce fine particles for suspension in water. In every case in which the particles were less than about 2 μm in diameter, they jiggled in the same way as those in the pollen grains of *Clarkia pulchella*.

THE NATURE OF BROWNIAN MOTION

Brownian motion is not specific to life. It arises from a perpetual dynamic energy of nature that permeates all of what Brown called the "molecules" of matter. For the remainder of the 19th century physicists debated the cause and nature of the phenomenon. In 1905, Albert Einstein published his theories of special relativity and of the quantum nature of light, but he was awarded his doctoral degree for showing how Brownian motion is related to the macroscopic phenomenon of diffusion. Four years later, Jean Perrin experimentally confirmed Einstein's equations for Brownian motion and showed that Brownian motion provided a means for calculating the Avogadro number, the number of molecules in a mole of gas or other substance. With the work of Einstein and Perrin, Brownian motion became part of the foundation of the new atomic-kinetic theory of matter.

The idea of atoms in motion goes back to the ancient Greeks, though neither Aristotle nor the Stoic philosophers nor later Descartes believed that matter is made of discrete atoms. The atom was reborn in more modern form at the beginning of the

19th century with the work of John Dalton, who introduced atoms and molecules to explain the properties of gases and the discrete combining proportions of chemicals in reactions. The particles observed by Brown were about the size of bacterial cells, far larger than what we call molecules today, but they were small enough to feel the impact of the water molecules in which they were suspended, being jolted this way and that when invisible but rapidly moving water molecules collided with them. Throughout all nature, including our own bodies, this ceaseless molecular motion abounds. It causes the oxygen in our lungs to diffuse into our blood, and from the blood to diffuse again to the billions of individual cells in our bodies. It drives water and minerals into the roots of plants and carbon dioxide into their leaves. Without this silent and unseen motion, all life would cease.

Where does this motion really come from? No one can say, anymore than we can say where matter comes from. Call it temperature, call it entropy, or call it thermal energy. By whatever name, it remains one of the fundamental mysteries of the universe in which we live. But one thing is sure: without its presence throughout nature, we would not live at all.

DIFFUSION AND THE UNSEEN ENERGY

As the studies of Jean Perrin and Albert Einstein made clear, the phenomena of Brownian motion and diffusion are intimately related. If a volume of solution contains regions of differing concentrations of particles, Brownian motion will cause the particles to spread from regions of higher concentration to those of lower concentration, and this movement of a quantity of individual particles or molecules down a concentration gradient is called *diffusion*. There are two ways for material to be moved from A to B: either as a massive assemblage of particles held together by cohesion and driven as one thing by an external force or pressure, or by diffusion of individual particles driven by their own internal kinetic energy to seek a uniform distribution in space. If we want to empty a pan of water, we can pour it out all at once (mass movement), or we can wait for it to evaporate (diffusion). The circulation of blood in our bodies is mass movement driven by the force of the heartbeat. The movement of oxygen from the blood vessels to the body cells (and to the mitochondria within the cells) is by diffusion.

Across large distances, diffusion is slow compared to mass movement, but it can be fast across short distances, particularly if surfaces of exchange are large. The living tissues of our bodies contain huge numbers of tiny blood capillaries that bring the blood close to every cell and create in sum an enormous surface area across which molecules can diffuse, so that all the oxygen, for example, that is carried long distances (from the lungs to the other organs) by the circulation is carried at equal rates across short distances by diffusion (from air to blood in the lungs and from blood to the various body cells).

Diffusion is dependent upon three factors: (1) the kinetic energy of the individual molecules (we will use the term "molecule" to include not only true molecules made up of multiple atoms, but atoms and charged ions as well); (2) a heterogeneity in the distribution of the molecules, so that the molecules are more concentrated in one region than another; and (3) a surface between the heterogeneous regions that is permeable to the individual molecules.

Diffusion and the Kinetic Energy of Molecules

The kinetic energy (KE) of the molecules is proportional to their mass (m) and to the square of their velocity (v). Written as an algebraic equation, and letting k be the coefficient of proportionality between KE and mv^2, we have $KE = kmv^2$, but, by convention of the units usually used for measuring mechanical energy, mass, and velocity, the coefficient k is $\frac{1}{2}$, so that this becomes $KE = \frac{1}{2}mv^2$.

Consider an 8 ounce glass of water. Externally, the water seems uniform, continuous, and placid. But, internally, about 7.6×10^{24} discrete molecules are bounding about, colliding with one another and continually changing places. The kinetic energy of any single molecule changes as it collides with other molecules, gaining velocity in one collision but losing it in another. In this maelstrom of collisions, however, the total kinetic energy within the water does not change for the gain in kinetic energy by one molecule is accompanied by an equal loss in another—assuming that no evaporation occurs at the water surface and ignoring momentary exchanges between strictly translational kinetic energy and energies of rotation or vibration in the water molecules.

Because of the continual gains and losses of kinetic energy among individual molecules, kinetic energies (and the velocities associated with them) are spread across a range of values for the entire population of molecules, but there is an average value for the population. Remarkably, that average energy is linearly proportional to the temperature of the water: the hotter the water, the greater the kinetic energy. Thus, mv^2 is proportional to temperature, and (assuming we are considering only one kind of molecule, so that mass is constant) velocity is proportional to the square root of temperature. The relationship between temperature and molecular kinetic energy is of immense importance to the chemical and physical activities of nature. The laundry on the line dries much more rapidly on a warm day than a cold one. Life itself is nearly confined to the molecular energies that range between the freezing and boiling points of water.

If we mix into the glass of water a bit of sugar, the sugar molecules are bigger than the water molecules, but they, too, join in the dance of collisions, and, by another remarkable fact, the larger molecules assume the same average kinetic energy as the smaller ones. As the US Constitution granted large and small states an equal number of senators, so nature and her physical laws have granted large and small molecules an equal share of kinetic energy at any temperature. The average value of mv^2 is the same for the sugar as for the water. But since the mass of the sugar molecule is greater than that of the water molecule, its velocity is less than that of the water. For a given temperature, the velocity of a molecule varies inversely with the square root of its mass. Because their velocities are less, bigger molecules diffuse more slowly than smaller molecules.

Diffusion and Concentration Differences

Heterogeneity is a key to all macroscopic activity in nature. If everything were the same in our world (the same temperature, the same chemical composition, the same pressure or voltage or density, etc.), nothing that we can directly perceive would ever happen. Add sugar to a cup of coffee, but forget to stir it, so that most of the sugar settles to the bottom. Initially, the coffee, sipped from the top, does not taste sweet, but over time it become sweeter without any help from us. To be honest, in this case most of the movement of sugar from bottom to top is by thermal convection currents rather than by simple diffusion, but if we could isolate the cup from any vibrations or changes in temperature, the sugar would (very slowly!) move by diffusion alone until it was evenly distributed throughout the coffee. If we had stirred the sugar up at the beginning, so that it was evenly distributed throughout the coffee, no change in sweetness would be detected over time. The transport of material by diffusion requires a concentration difference between two regions.

Consider a boundary between two regions, A and B, where A contains a great many molecules of a substance, and B contains very few. Molecules are moving about in all directions on both sides of the boundary, but because there are more molecules in A to move toward B than there are in B to move toward A, the result is a net flux of molecules across the boundary from A to B. Across the boundary from A to B is a *concentration gradient* (defined as the change in concentration with distance, $\Delta C / \Delta x$), and the greater the concentration gradient the faster is the rate of diffusion. Like marbles rolling down an incline, the steeper the incline, the faster the movement. Without an incline—with no concentration gradient—there is no diffusion.

The rate of transport by diffusion is also proportional to the area of the boundary across which the molecules can move. The movement across one part of a boundary adds to that in another part, so that the greater the total area for diffusion, the faster the diffusion occurs. Hence, rate of diffusion (in mass per time, $\Delta m / \Delta t$) is proportional both to the concentration gradient ($\Delta C / \Delta x$) and to the area (A) of contact. The proportionality coefficient relating these factors is called the *diffusion coefficient* (D), so that we have:

$$\Delta m / \Delta t = -DA(\Delta C / \Delta x)$$

This relationship is known as *Fick's first law of diffusion*, after Adolph Fick who proposed it in 1855, long before the work of Einstein and Perrin on Brownian motion. The minus sign signifies that movement of material is in the direction of decreasing concentration. The diffusion coefficient incorporates three essential considerations because it is influenced by the nature of the diffusing molecule (especially its size), the nature of the material through which it is diffusing, and the temperature (since temperature determines the amount of kinetic energy the diffusing molecules have). For example, glucose diffusing in water at $25°$C has a diffusion coefficient that is different from that of a bigger or smaller molecule at that same temperature and different from its own diffusion coefficient at some other temperature or in some other medium.

As an example of how this works, imagine a cube, 1 centimeter on a side, separating a solution of glucose on one side and pure water on the other. The glucose solution is 1 molar in concentration (1 mole per liter, where a mole equals 6×10^{23} molecules of glucose, or 180 grams). Suppose that, across the cube, the glucose concentration drops off linearly from 1 molar to zero, so that we have a constant concentration gradient across the cube. At $25°$C, the diffusion coefficient of glucose in water is about 6×10^{-6} cm^{2}sec^{-1}, which translates for this concentration gradient and area to give a very slow diffusion of glucose from one side of the cube to the other. It would take a little over 5 years for 1 mole of glucose to diffuse across. But if we have 1 square meter of surface available for the diffusion instead of the paltry 1 square centimeter (the distance from one side to the other remaining 1 centimeter), the diffusion rate increases 10,000-fold, and a mole of glucose would pass across in a little less than 5 hours. If we further reduce the distance separating the two sides from 1 cm to 10 μm, the rate of diffusion increases another 1,000-fold, and a mole can pass in about 16 seconds.

This example assumes that the concentration of glucose is maintained at its original level on the supply side in spite of losses and that the recipient side remains near zero in concentration in spite of gains. Such a "steady-state" is not unrealistic in our living bodies, where the blood vessels continually deliver oxygen, for example, to within 10 μm of our body cells, and the cells continually use up the oxygen that diffuses to them across the intervening space.

If such a steady-state is not maintained by external factors (such as blood flow and cellular utilization), then the concentration gradient that drives the diffusion will gradually be destroyed by its own action as material is removed from A and delivered to B, causing the concentration of A to fall and that of B to rise. The rate of diffusion diminishes with the decrease in concentration gradient, and, over time, the two regions head toward equality in their concentrations, creating a homogeneous system at equilibrium. Like a hot body in contact with a cold one, where entropy passes from the higher temperature to the lower, causing both to achieve a common temperature, the diffusing substance passes from high concentration to low until a common concentration is achieved. Activity requires a gradient, such as the concentration gradient or temperature gradient, and the result of the activity is the destruction of that gradient. In such simple systems, we start with heterogeneity (gradients) and end with homogeneity (uniformity). The inclined plane becomes a level playing field.

SPREADING BY RANDOM WALK ALONG A CONCENTRATION GRADIENT

Let us revert to the mysterious motion described by Brown. Brown was looking at particles attempting to diffuse, to move from high concentration to low if such concentration differences existed. Not that the particles want to move, yet something makes them move. It is common to say that they move because they are struck by rapidly moving but invisible water molecules around them. Yet nature is interrelational, an action received is a reaction given, and, as the water molecules hit the large particles, so the large particles hit the water molecules. The large particles are just as much self-moving as are the small molecules, though they move with lesser velocities because they are more massive. Large or small, they all have the same average kinetic energies, expressing a common primordial energy that we associate with temperature.

Einstein thought about how Brownian particles should behave if their motion is random—that is, if each particle is equally likely to move to the right or left, up or down, backward or forward in the three dimensions of space, independently of the other particles and independent of its own previous motions. To simplify the model, consider just one dimension instead of three, so that molecules move either right or left along a line, and suppose that their motions are always in very short steps of constant length. After a given step, the molecule may continue straight ahead in the next step, or it may reverse direction because of a collision with another molecule.

To see how this "random walk" would work, we will let 16 molecules all start at the zero point of a line and require that each take one step to the right of the origin (positive) or to the left (negative). On average (in a great many such trials), eight should move to a –1 position and eight to a +1 position. In one sense, nothing has happened, for the mean position of all the molecules is still zero. Yet, in another sense, something very fundamental has happened, for the molecules have in fact moved outward from the origin.

With step 2 (again taking an average for a great many trials), four of the molecules at the –1 position move outward to the –2 position, and four move back to the origin. The same occurs on the positive side of the origin, so that, at the end of step two (on average!—and let us stop saying so), we find four molecules at –2, eight molecules at 0, and four molecules at +2. While no progress has been made in changing the average position of all molecules (the mean position for all is still at zero), the molecules are beginning to penetrate the space outward from the origin and to separate themselves from one another.

At the end of the third step of this random walk, two molecules are at +3, six are at +1, six are at –1, and two are at –3. And at the end of the fourth step, one molecule is at +4, four are at +2, six are at 0, four at –2, and one at –4.

After four steps, the mean position of the molecules is still at 0, but we see that they are moving outward in both directions. This progress over time (each step being associated with a duration of time) can be expressed quantitatively if we take the squares of the positions of the molecules, sum up the squares, and then take the average of those squares. Because negative numbers become positive when they are squared, the sum of squares increases with time even though the sum itself remains at zero.

Letting N be the total number of molecules, and n be the number at each distance, x, from the origin, the sums of distances and distances squared can be represented by $\sum nx$ and $\sum nx^2$, and the respective averages by those sums divided by N. We have then:

Step 1

$$(\sum nx) / N = \left[8(-1) + 8(1) \right] / 16 = 0$$
$$(\sum nx^2) / N = \left[8(1) + 8(1) \right] / 16 = 1$$
$$\text{root mean square distance} = \sqrt{1}$$

Step 2

$$(\sum nx) / N = \left[4(-2) + 8(0) + 4(2) \right] / 16 = 0$$
$$(\sum nx^2) / N = \left[4(4) + 8(0) + 4(4) \right] / 16 = 2$$
$$\text{root mean square distance} = \sqrt{2}$$

Step 3

$$(\sum nx)/N = \left[2(-3) + 6(-1) + 6(1) + 2(3)\right]/16 = 0$$
$$(\sum nx^2)/N = \left[2(9) + 6(1) + 6(1) + 2(9)\right]/16 = 3$$
$$\text{root mean square distance } = \sqrt{3}$$

Step 4

$$(\sum nx)/N = \left[1(-4) + 4(-2) + 6(0) + 4(2) + 1(4)\right]/16 = 0$$
$$(\sum nx^2)/N = \left[1(16) + 4(4) + 6(0) + 4(4) + 1(16)\right]/16 = 4$$
$$\text{root mean square distance } = \sqrt{4}$$

A pattern is certainly emerging. A little algebra could prove that the pattern will go on for as many steps as we wish, but we will take it as fairly obvious. The average distance from the origin will always be zero, but the average square of the distances will increase linearly with the number of steps. But what exactly is the square of a distance? It is something akin to an area (a length squared), and we are interested in length itself. So the square root of the average square is taken to yield a "root mean square" distance, which is truly a distance and nothing but a distance.

During the random walk of a great many Brownian particles, the absolute distances of the particles from the origin are spread out from zero to a maximum, which is the product of the number of steps times the step length. But the population of molecules will have a specific root mean square distance, and that distance increases with the square root of the number of steps times the step length—or with the square root of time, assuming the frequency of steps is constant with time. Thus, the random walk model predicts that Brownian particles will spread with time, and the spread, as described by the root mean square distance, will increase with the square root of time. This behavior was predicted by Einstein and was experimentally observed by Jean Perrin.

THE MOTION THAT NEVER STOPS

This spreading phenomenon—the dissolving of sugar in our coffee, the drying of clothes on the line, the scent in the air from newly mown grass—is such a common part of our lives that we give it scarcely any thought. But behind the ordinary lies the fundamental, and in the dispersion of molecules we have one of the most elemental features of nature and our existence. What do Brownian motion and the diffusion associated with it have to tell us about the way nature works?

As Perrin noted, the most extraordinary thing about Brownian motion is that it never stops. In our macroscopic world, as Galileo and Newton taught us, motion is just as natural as rest (it requires no force to explain it)—but only if there is nothing resisting the motion. The planets can orbit eternally (almost) around the sun because outer space offers no resistance to their motion. But is there no resistance to the motion of molecules or particles through such a fluid as water?

Viscous forces impede a large object moving through water, and the object soon comes to rest unless it continues to be impelled by a driving force. Einstein assumed that microscopic objects, including Brownian particles or even true molecules, are also subject to viscous forces. The collisions of one molecule with another are resisting forces, and a Brownian particle moving at one instant in one direction is in another moment moving in a different way, and these changes in velocity mean that it is being subjected to external forces. But, in spite of viscous forces that resist its motion, the Brownian particle never stops moving. So, in the view of Einstein (if I interpret his writings correctly), each particle is subjected to a never-ending driving force that balances the never-ending viscous force and keeps it in motion forever. What is this driving force? It is the force of thermal agitation, which constantly assures to all particles a kinetic energy. What impedes a particle in one direction pushes it in another, and the motion, in one direction or another, goes on forever. At the microscopic level, perpetual motion can exist in a resistant medium. If any motion were lost via an inelastic collision, there would be nowhere for the kinetic energy to go except to increase the thermal agitation, which would increase to an equal measure the energy being depleted—which is to say, in effect, that all microscopic collisions must be elastic. That is one fundamental lesson that Brownian motion seems to teach us.

Brownian motion also suggests that, at a microscopic level, classical boundaries between mechanical work and thermal energy (heat) disappear. In 1824, 3 years before Brown's discovery of microscopic particle motion, Sadi Carnot deduced that the production of mechanical work by a steam engine or other fire-driven contrivance requires not only a hot source but a cold sink. A steam engine can achieve nothing without a temperature difference, just as a water wheel can produce no mechanical work without a difference in height across which water can flow. Yet, in studying Brownian motion, Perrin noted that microscopic particles denser than water sometimes move momentarily upward against the force of gravity. In such moments, mechanical work is done at the expense of thermal energy, without the presence of any temperature gradient. Conversely, when the particle moves downward again, gravitational work is converted to thermal energy. When

Brownian motion of dense particles occurs up and down in a gravitational field, microscopic bundles of mechanical and thermal energy are being exchanged all the time.

But perhaps the most important lesson from Brownian motion is the simple one that materials of nature have a tendency to spread. In the 20th century, astronomers discovered that galaxies are receding from us, and cosmologists today theorize freely about our expanding universe. Although all nature ultimately is interrelated, one would not wish to claim a direct connection between the large-scale cosmological forces that cause our universe to expand and the microscopic thermal forces that sustain Brownian motion and diffusion. Yet the analogy should not be dismissed too lightly, for both phenomena portray a fundamental asymmetry relating to time, pronouncing that the past gives way to the future—that time has direction as well as duration. We would not expect the outer galaxies to turn about and start moving towards us and, if they did, we should certainly be alarmed! Nor would we expect thermal agitation to reverse the direction of diffusion and cause sugar that has evenly distributed itself in our coffee to come back together into a cube at the bottom of the cup. Galaxies disperse with time and so do molecules.

The phenomenon of diffusion—of dispersal of material through thermal agitation—is intimately related to entropy, since entropy is the extensive factor of thermal energy. At the microscopic level of Brownian motion and diffusion, thermal energy drives a process of change even though there is no temperature gradient—that is, equipotential thermal energy contributes to the free energy of a different sort of gradient— the gradient in this case of concentration across space. The result of the diffusion is the achievement of a uniform distribution of material (and the destruction of the concentration gradient). The entropy of the diffusion system increases in this process.

The diffusion process might be coupled to outside work—as in an electrical concentration cell or in the chemiosmotic synthesis of adenosine triphosphate (ATP) in the living cell (topics to be considered in the next two chapters). If the diffusion process is fully compensated (reversible), the environment of the diffusion process would lose as much entropy as the diffusion system itself gains, and the entropy of the world would remain constant. But, in any spontaneous process, the destruction of the concentration gradient is not fully compensated, and the entropy of the world (diffusion system plus environment) increases.

Diffusion is a model for all natural processes, for whenever anything happens in nature, a gradient of some kind is involved, and that gradient is diminished by the activity it drives. Even the transfer that occurs between a hot body and a cold one obeys this rule. As something disperses from high temperature to low, the temperature difference is gradually decreased, and, in the end, a common, middle temperature is reached. What disperses across a temperature gradient is entropy, and the communication of this entropy provides the microscopic kinetic energy of Brownian motion and diffusion.

It is common to speak of the dissipation of energy that accompanies all natural processes. The material that has diffused to uniform concentrations can diffuse no longer, nor can it reverse itself and undo its own dispersal. Something has been lost as something was achieved. In science, as in life more generally, we can look at ultimate meaning in any way we wish. But if one person insists that nature is running downhill and burning itself out, it is perfectly reasonable for another to point out that without the random motion seen by Brown and the diffusion that can accompany it, no life as we know it could exist on our Earth.

Chapter 7

Chemical Energy

Chemical reactions, like all other natural changes, require the existence of gradients and involve the conversion of free energy, which can do work, into other forms of free energy and/or into bound energy, which in its lack of gradients can do no further work. Equilibrium is reached when gradients and free energy differences have been reduced to zero.

But what is a chemical gradient or potential? Where are the waterfalls across which moles of chemical reagents flow and, in flowing, can do work? While the gradients of gravitational, mechanical, electrical, or thermal energy can be measured by relatively simple means (a thermometer will do for the latter!), the chemical potential is more complex, and the discovery of its secrets has been one of the great achievements of modern science. Three main methods have been used for the measurement of chemical potentials and free energy. These have involved (1) the measurement of heat given off or absorbed in chemical reactions, (2) the measurement of electrical potentials generated by certain reactions, and (3) the quantitative analysis of the concentrations of chemical reactants and end products when equilibrium has been reached.

THERMAL ENERGY IN CHEMICAL REACTIONS

Heat, or caloric as Lavoisier called it, seems to be almost a reagent in chemical reactions. It cannot be weighed, but, as Lavoisier and Laplace were the first to show, it can be accurately measured by calorimetric means, as by determining the amount of ice that is melted by a given amount of reaction. When coal is burned, it produces a measurable amount of caloric. When twice the coal is burned, twice the caloric is recorded. The caloric is so much a part of chemistry that Lavoisier listed it as one of the elements. The conservation of the elements in chemical reactions suggested also the conservation of caloric—except, as Robert Mayer and Hermann Helmholtz later saw in the animal body, caloric (heat) and work are bonded together in mutual dependence upon the same oxidative reactions so that work and heat must be in some sense equivalent. And so was born in the 1840s, as we have seen, the law of the conservation of energy.

But in 1840, 2 years before Mayer's initial paper, Germain Henri Hess (Swiss born, but working in Russia, 1802–50) published a conservation law, known today as *Hess's law*, regarding the way heats of chemical reactions relate consistently to one another. The laboratory chemist knows all too well that care must be taken in diluting concentrated sulfuric acid with water because so much heat is generated by the reaction between acid and water. If the dense acid is added slowly to the water, it will sink and distribute the heat throughout the vessel. If the water is added to the acid, it will float on the surface and boil. Many a laboratory coat has acidic holes spattered into it from the heat of this reaction. But Hess studied the reaction between water and sulfuric acid carefully, determining that sulfuric acid forms definite hydrates with water, taking on successively one, two, and then four molecules of water for each sulfuric acid molecule. Hess found that the sum of the heats produced in the successive hydrations equaled the heat produced if the full hydration were carried out all at once. This he called the *law of constant heat summation*, and he recognized the significance the law would have in predicting heats of reactions that had not yet been directly measured. He understood that if a hydrocarbon were burned, it would give off less heat than an equal amount of hydrogen and carbon burned separately because some heat would already have been given off when the hydrogen and carbon combined to form the hydrocarbon:

> The sum of the heat corresponding to a certain amount of water and carbonic acid which we suppose arises from combustion, *being constant*, it is evident that if hydrogen is found previously combined with carbon, this combination cannot have occurred without evolution of heat; this amount, already eliminated, cannot be recovered in the quantity evolved by the defined combustion. There results in practice this simple rule: that *a combustible compound always evolves less heat than do its elements taken separately*.[1]

Hess hoped that chemical equations would in the future express the amount of heat involved as well as the number of atoms, and he suggested that those amounts might, like the atoms, occur

---

1. Germain Henry Hess, Comptes rendus 10, 759–763 (1840). Cited in Henry M. Leicester, "Germain Henry Hess and the Foundations of Thermochemistry," *Journal of Chemical Education*, November 1951, p. 582. Emphasis in the original.

in definite proportions and might shed light on the nature of chemical affinity—on how things bond with one another:

> I have the full conviction that we will have a precise idea of chemical phenomena only when we succeed in indicating in our formulas the ratios of heat relations as we indicate today the relative numbers of ponderable atoms; at least thermo-chemistry promises to disclose the still secret laws of affinity.[2]

The study of thermo-chemistry was continued by Julius Thomsen (1826-1909) in Denmark, and by Marcellin Berthelot (1827-1907) in France. Both Thomsen and Berthelot concluded that the heat of reaction was a quantitative measure of the affinity of the reacting substances. But Berthelot knew that some reactions absorbed rather than produced heat and coined the words "endothermic" and "exothermic" for the two cases.

Heat production alone does not in fact tell us whether and to what extent chemicals might react with one another. When nitre (potassium nitrate, KNO_3) is dissolved in water, the solution becomes cold, but when it is mixed with carbon and sulfur it forms gunpowder and becomes explosive. The reaction of nitre with water is endothermic, and with carbon and sulfur it is highly exothermic, yet both these reactions proceed spontaneously.

Heat is produced in chemical reactions by the formation of bonds between atoms and molecules. Chemical bonds—whether they be the strong covalent bonds of shared electrons or weaker ionic, hydrogen, van der Waals, or hydrophobic bonds—represent attractive forces between atoms, and the forces bringing those atoms together can do work or produce thermal energy in the process. The formation of a bond liberates energy, and the breaking

2. Ibid.

of a bond requires that energy be put back in— like an exchange between kinetic and potential energy in macroscopic mechanics, as when a stretched spring can do work in pulling inward and then requires work to be stretched out again. To a first approximation, a chemical reaction is a trading of weak bonds for stronger ones, with the liberation of energy that can express itself either in work or in heat. But, in addition to the attractive electrical forces that pull atoms together, nature has a repulsive force always present in chemical reactions. The repulsive force is thermal. We have seen it in Brownian motion and in diffusion down a concentration gradient. It is entropy in action, the agitation that goes with warmth, with high temperature. In a sense, Aristotle was right to say that nature abhors a vacuum. Give matter empty space and it will try to occupy it. Everything has an "escaping tendency" and is repelled by thermal energy from being where it is. At high enough temperature, all chemical bonds are broken. Life as we know it can live only in a very restricted temperature range, not only because liquid water is confined (at our atmospheric pressure) to 100 centigrade degrees, but because the weak bonds that hold the three-dimensional structures of proteins, nucleic acids, and other macromolecules together are disrupted by modest temperature increases.

To the heats of reaction that express bonding tendencies, we must add the tendencies of components to break away and spread out from one another, to mix with other components, to occupy the total volume available, to expand the volume in the case of gases: to decrease the free energy gradients of concentration by distributing all components uniformly. The free energy available from a chemical reaction is a sum of two factors reflecting the competing forces of specific attraction and nonspecific repulsion.

Many spontaneous chemical reactions are endothermic, causing the temperature of the reaction mixture to fall rather than rise. The drop in temperature becomes part of a new equilibrium if the reaction vessel is thermally insulated from the environment. Examples of such reactions are the dissolving of potassium chloride (KCl) or sodium nitrate ($NaNO_3$) in water. The tendency of the salt ions to spread throughout the solution overcomes the fact that the interionic bonds within the original salt crystals are somewhat stronger than the bonds between the ions and water. Here is a case where stronger bonds are traded for weaker ones, driven by thermal agitation and the pursuit of uniform distribution of components. It is similar to the evaporation of water vapor from water, where the spreading (escaping) tendency of the water molecules overcomes the intermolecular bonds in liquid water. The evaporation takes entropy (and thermal energy) from the water, causing the water to cool—an endothermic reaction that is a vital part of temperature regulation in warm-blooded animals.

INTERNAL ENERGY AND ENTHALPY

The fact that heat emerges from exothermic chemical reactions implies, in view of the conservation of energy, that energy was in the chemicals before they reacted. To preserve the conservation of energy, Rudolf Clausius introduced the concept of internal energy, which he signified by the letter U, but which is more commonly today symbolized by E. Internal energy is considered to be a state variable.

Chemical states of equilibrium are defined in terms of attributes, called *state variables*, whose values are determined for any particular equilibrium state. Like the coordinates defining a point on a graph or in space—like the coordinates of latitude and longitude that specify a unique position on the Earth's surface—state variables define a unique state. When a system changes to some other condition, and then changes back to its original state, a state variable must return to its original value, having undergone an overall change of zero, no matter what pathway of changes the system underwent. The difference between two states can be described by the differences in their state variables, just as the different positions of London and New York can be described by the differences in their latitudes and longitudes.

Temperature, pressure, volume, entropy, and mass are state variables, all of which can be measured on an absolute scale. Volume and mass have obvious zero points where they disappear. With the invention of vacuum pumps and barometers, ambient pressure can be seen to have an absolute zero. With the discovery of a point on the thermal scale where extrapolated gas volumes disappeared, temperature acquired an absolute zero, as did entropy, too, with the postulate of the third law of thermodynamics.

The state variables of internal energy, enthalpy, and free energy are somewhat different as we cannot know their absolute values. We know them only by differences between two states, by the heat and work that goes in or comes out in changing from one state to another—rather like living in hill country, without any access to the sea: we can measure the difference in elevation between a hill and a valley, but we cannot know the absolute elevation of either. Such state variables might be called *deltoid state variables* to remind us that we measure them by differences between one state and another, assigning for convenience one "standard state" a value of zero.

The internal energy, E, of a chemical system can be changed by an interaction with its environment that involves either heat, q, or work, w:

$$\Delta E = E_2 - E_1 = q - w$$

In words, the change in internal energy (ΔE) is the difference between the final and initial internal energies $(E_2 - E_1)$, which in turn equals the positive heat received by the system minus the positive work given by the system. Clausius used this sign convention $(+q$ but $-w)$ because it describes a steam engine receiving heat but giving work, and most textbooks follow his example, although some authors prefer to use $+w$, which has the effect of making all inputs to the system positive and all outputs negative. Those who do not find either convention confusing at times are surely among a very lucky few.

Energy received as work by a chemical system is strikingly illustrated by compressing a gas, which is warmed by compression as surely as if it had been warmed by heating. And, likewise, energy is given up as work when a gas expands against a piston, cooling itself in the process. But a chemical system can also do work in creating an electrical gradient, in creating a concentration gradient, in synthesizing complex molecules, in generating light, or even in producing motion (as in the contraction of muscles).

If all work is excluded as a means of exchange, then changes in internal energy can only happen if entropy (and with it heat) flows into or out of the system. If that happens, then $\Delta E = q$. If the heat exchange occurs at constant volume, then $\Delta E = q_v = C_v \Delta T$, where C_v is the heat capacity at constant volume. If the exchange is at constant pressure, we might think that $\Delta E = q_p = C_p \Delta T$, where C_p is the heat capacity at constant pressure. However, at constant pressure, the volume of the system is likely to change, which introduces a work term of $P\Delta V$, so that $\Delta E = q_p - P\Delta V$, or $\Delta E + P\Delta V = q_p$.

Since most chemical reactions are carried out in an open vessel at constant (atmospheric) pressure, it is convenient to construct a new state variable that allows for unseen work done by changes in volume. This state variable is called *enthalpy* (some authors have called it the *heat function* or *heat content*) and is given the symbol H. The only difference between enthalpy, H, and internal energy, E, is that a PV energy term is added to E to give H:

$$H = E + PV$$

What is this PV energy term? It is the volume of the chemical component times its (atmospheric) pressure. It is a small factor for solids and liquids, but much larger for gases: for example, 1 mole (18 grams) of liquid H_2O has a volume of only .018 liter at $25°C$, but 1 mole (2 grams) of H_2 gas has a volume of 24.4 liters, so that the PV term for the H_2 gas is 1,356 times bigger than for the H_2O liquid. In terms of calories, PV is about 592 calories for the H_2 gas but only 0.44 calories for the H_2O liquid.

When enthalpy changes, we have:

$$\Delta H = \Delta E + \Delta(PV) = \Delta E + P\Delta V + V\Delta P$$

But for reactions run at constant pressure, the last term, which represents the energy of a pressure gradient, does not occur since ΔP is zero. Thus, at constant pressure:

$$\Delta H = \Delta E + P\Delta V$$

Previously, we saw that heat added at constant pressure, q_p, is equal to the right-hand side of this expression. Hence, $\Delta H = q_p = C_p \Delta T$

In summary, when heat is added to a system at constant volume, the internal energy increases by the amount of heat added. When heat is added at constant pressure, the enthalpy increases by the amount of heat added. For a given temperature increase, the increase in enthalpy (and the q_p involved) is usually greater than the increase in internal energy (and the q_v involved), since the temperature increase usually causes an increase in volume and hence a positive $P\Delta V$ term (which represents work done by the system on the environment).

Unlike pressure, volume, temperature, and entropy, the state variables of internal energy and enthalpy cannot be known on an absolute scale for any chemical substance. For convenience, enthalpy is symbolized by H, or by H° if standard conditions of temperature (usually 298° K) and pressure (1 atmosphere) are indicated. But what H really represents is what the substance has undergone thermally to become what it is, for H indicates the extent of heat output (or occasionally input) that has accompanied the formation of the substance from its elements. Full disclosure labels this as ΔH_f, the heat of formation of a substance, or ΔH_f°, the heat of formation under standard conditions; that is, H and H° are convenient shorthand abbreviations for ΔH_f and ΔH_f°. For example, water has a ΔH_f° (or H° for short) of −68.3 kcal/mole. This means that 68.3 kcal. of heat is released when 1 mole of H_2 combines with a ½ mole of O_2 to form one mole of H_2O. The elements that combine to form water—hydrogen and oxygen—are assigned ΔH_f° values of zero when they are in their most stable forms, namely, the diatomic molecules, H_2 and O_2. Assigning zero enthalpy values for the elements is entirely arbitrary but very helpful. It is like assigning zero longitude to Greenwich, except that there is only one Greenwich and there are many elements.

Standard values for the enthalpy of formation have been tabulated for many substances. From these values, the change in enthalpy for many chemical reactions can be predicted for the change in enthalpy, ΔH, is the sum of the ΔH_f° values for the end products of the reaction, minus the sum of the ΔH_f° values for the reactants:

$$\Delta H^\circ = \Sigma \Delta H_f^\circ \ (\text{products}) - \Sigma \Delta H_f^\circ \ (\text{reactants})$$

For example, the oxidation of acetaldehyde to acetic acid can be represented as:

$$CH_3CHO + \tfrac{1}{2}\,O_2 \rightarrow CH_3COOH$$

The ΔH_f° values in kcal/mole are:

Acetaldehyde −49.9
Oxygen   0   (by definition)
Acetic acid   −116.3

Then, the ΔH for the reaction is:

$$(-116.3) - (-49.9 + 0) = -116.3 + 49.9 = -56.4 \text{ kcal/mole}$$

Hess's law can be used to calculate the ΔH_f° for a substance such as CH_4 (methane) that cannot be easily formed from its elements in a calorimeter:

$$C + 2H_2 \rightarrow CH_4$$

The oxidation of all three of the substances involved in this reaction can be carried out in a calorimeter, yielding the following ΔH values in kcal:

1. $CH_4 + 2O_2 \rightarrow CO_2 + 2H_2O$ $\Delta H = -191.8$
2. $C + O_2 \rightarrow CO_2$ $\Delta H = -94.1$
3. $2H_2 + O_2 \rightarrow 2H_2O$ $\Delta H = -115.6$

If the first reaction is reversed and added to the other two, the net result is the formation of methane from carbon and hydrogen, with a calculated ΔH_f° of $(+191.8) - 94.1 - 115.6 = -17.9$ kcal/mole.

THE EFFECT OF TEMPERATURE ON ΔH

We have noted that the ΔH° for a reaction equals the sum of the ΔH_f° values for the products minus the sum of the ΔH_f° values for the reactants. But that little superscript circle beside the ΔH is a reminder that we are talking about standard conditions of pressure and temperature (usually $25^{\circ}C$ or $298.15^{\circ}K$). What happens to ΔH if the reaction is carried out at a different temperature? The change in ΔH with temperature depends only on how much heat has to be added (or subtracted) from reactants and end products to alter their temperature from $298^{\circ}K$ to the temperature at which the reaction is run. This heat is the *heat capacity* C_p for each substance times the change in temperature, or, if the heat capacity is not constant, it is the integral $\int_{298}^{T} C_p dT$, where T is the reaction temperature. We add this amount of heat to each reactant and to each end product (using the appropriate value of C_p for each). In effect, we will have changed the heat of formation of each substance from its standard value at $298^{\circ}K$ to the value it has at temperature T, that is, we will have calculated its ΔH_f at temperature T from its ΔH_f°. The new heat of reaction for converting reactants to products is then:

$$\Delta H = \Sigma \Delta H_f (\text{products}) - \Sigma \Delta H_f (\text{reactants})$$

This new ΔH is related to the standard ΔH° by the difference between the sum of the heat capacities of the products and the sum of the heat capacities of the reactants:

$$\Delta H = \Delta H^{\circ} + \int_{298}^{T} C_p (\text{products}) dT - \int_{298}^{T} C_p (\text{reactants}) dT$$

If products and reactants all happen to have the same heat capacities, then a change in temperature has no effect on the heat of reaction; that is, $\Delta H = \Delta H°$.

In summary, the heat of a reaction depends on the nature of the chemicals involved (as reflected in their respective standard heats of formation from their elements), on the temperature of the reaction, and, of course, on the amount of reaction carried out (which depends on the amount of the limiting reactant involved). But the heat of reaction does not depend significantly on the *concentrations* of the reactants and products. This is quite unlike (as we shall see) the free energy change for a reaction, which depends also on concentrations.

THE RELATION OF INTERNAL ENERGY TO FREE ENERGY

Except in the realm of electrochemistry, where chemical reactions create voltages that can drive a flow of electricity to run electric motors and so forth, the chemist is not aware that any work is being done when one chemical reacts with another. Except for the work of expansion (if it occurs) against the atmospheric pressure in the laboratory, heat seems to be the only energetic expression of the reaction, and, since thermal energy can only sparingly produce work (given the limited temperature range available in the laboratory), it is tempting to suppose that the internal energy is largely unavailable for doing work. But this is not the case, as is strikingly demonstrated by the generation of electricity in electrochemical reactions. The internal energy of reacting chemicals is mostly free energy, as it represents a real heterogeneity, a real gradient that can do work until it destroys itself in the uniformity of equilibrium. Much of the energy given off as heat in chemical reactions could have emerged as work if only another kind of gradient, such as that witnessed in electrochemistry, could be coupled to the reactions. Living organisms have been utilizing such couplings for hundreds of millions of years in the biochemical synthesis of complex molecules, in physical transport of ions and nutrients, and in generating mechanical force and motion.

One part of the internal energy is unable to do work except under particular circumstances. This is the *thermal energy, TS,* the product of entropy times the temperature of the system. This energy is equipotential (bound); that is, it contains no gradients unless either temperature or entropy is allowed to change. If reactions are carried out reversibly at constant temperature, the temperature does not change, and TS can affect maximum work only if the reaction involves changes in entropy associated, for example, with changes in volume or mixing.

The internal energy consists of two parts, free energy, F, and bound energy, TS:

$$E = F + TS$$
$$F = E - TS$$

The free energy, F, is called the *Helmholtz free energy*, and it is the maximum work that the system can do at constant volume. More commonly used is the *Gibbs free energy, G*, which is the maximum work that can be done at constant pressure. Henceforth, in referring to chemical free energy, we will mean the Gibbs free energy. Since $G = F + PV$ and $H = E + PV$:

$$G = H - TS$$

This equation says, in effect, that free energy is equal to the total energy minus the bound energy.

THE GIBBS FREE ENERGY OF FORMATION OF A SUBSTANCE

Just as for the internal energy or the enthalpy, there is no way to measure the absolute value of the free energy of a substance. Instead, the difference in free energy between the substance and its elements—the free energy of formation, ΔG_f°—can be calculated from the ΔH_f° for the substance and values of S° for the substance and its elements:

$$\Delta G_f^\circ = \Delta H_f^\circ - T\Delta S^\circ$$

where ΔS° equals the difference between the entropy of the substance and the sum of the entropies of its elements. As an example, let us calculate the ΔG_f° for methane. We have already calculated the ΔH_f° for methane:

$$C + 2H_2 \rightarrow CH_4 \quad \Delta H_f^\circ = -17.88 \, \text{kcal} / \text{mole}$$

The S° values for CH_4 and its elements, C and H, are:

CH$_4$ 44.46 cal/mole
C   1.86
H$_2$ 31.23

$$\Delta S^\circ = 44.46 - \big(1.86 + 2(31.23)\big) = 44.46 - 63.82$$
$$= -19.36 \, \text{cal} / \text{mole}$$

$$T\Delta S^\circ = (298)(-19.36) = -5.77 \, \text{kcal} / \text{mole}$$

$$\Delta G_f^\circ = \Delta H_f^\circ - T\Delta S^\circ = -17.88 - (-5.77) = -12.11 \text{ kcal / mole}$$

In this particular case, the ΔG_f° is less negative than the ΔH_f°. What does this mean? Because the ΔS° value is negative—that is, the entropy of the product methane is less than the sum of the entropies of the reactants carbon and hydrogen—the subtraction of $T\Delta S^\circ$ is in effect an addition of a positive term to ΔH_f°, thus lowering its negativity. If the free energy change is diminished by such a decrease in entropy, then enthalpy has to be discharged as heat rather than work; that is, an amount of heat equal to $\left(-T\Delta S^\circ\right)$ is conveyed from the chemical reaction system to the environment. The reaction is assumed to be reversible, so that the overall entropy change of the system and its environment is zero. Hence the entropy decrease of the chemical system must be exactly balanced by an entropy increase of the environment. Similarly, in a reversible reaction in which the ΔG_f° is more negative than the ΔH_f° because the change in entropy is positive, then entropy (and heat) flows into the reaction system from its environment, so that the entropy increase of the reaction system is balanced by an entropy decrease in the environment

FREE ENERGY OF A CHEMICAL REACTION

Just as the ΔH° for a chemical reaction is the difference between the sum of the ΔH_f° values for the end products and the sum of the ΔH_f° values for the reactants, so, by the same reasoning, the ΔG° for a reaction can be calculated:

$$\Delta G^\circ = \Sigma\Delta G_f^\circ \left(\text{products}\right) - \Sigma\Delta G_f^\circ \left(\text{reactants}\right)$$

The ΔG for a chemical reaction depends on the temperature and pressure, as does the $\Delta G°$. The ΔG (unlike the $\Delta G°$) also depends, in a highly significant way, on the concentrations of the reactants and end products. The inclusion of a little circle (°) to the right of the ΔG indicates standard concentrations (usually 1 molar) for the reactants and products involved in the reaction. Thus, $\Delta G°$ is the free energy change for a reaction in which all the reactants and products are present at 1 molar concentration. If the reaction involves gases, the standard concentration for the gases will usually be a pressure of 1 atmosphere.

In biochemistry, where many reactions involve H^+ or OH^- ions, a standard condition of 1 molar for those ions would be far removed from the usual conditions, and the $\Delta G°$ is often defined in the usual way *except* that the concentration of H^+ is taken to be 10^{-7} molar instead of 1 molar. If that is the case, the $\Delta G°$ becomes $\Delta G°{}'$, with the prime mark to warn the reader that the standard condition includes a pH value of 7. Likewise, $\Delta G'$ refers to the free energy of a reaction system when the standard concentrations include a pH of 7.

The free energy of formation of a substance, which is a deltoid function—one that represents the difference in free energy between a substance and the elements of which it is made—will subsequently be referred to simply as the free energy of the substance and will be represented as $G°$ for the standard state or as G for a nonstandard state.

For the reaction:

$$A + B \rightarrow C + D$$

The free energy change is:

$$\Delta G = \left(G_C + G_D\right) - \left(G_A + G_B\right) \tag{7.1}$$

where G_A, G_B, G_C, G_D are the free energies of the reactants and products involved in the reaction. But the free energy of each substance is, as we have seen previously, a function of its concentration, increasing with greater concentration and decreasing with lesser concentration. For 1 mole of substance, the difference in free energy between two different states of concentration is given by:

$$G_1 - G_2 = RT \ln \frac{C_1}{C_2}$$

Let the first state be the substance at any concentration, C, and the second state be specifically the standard state with a concentration of C°. We have then for the difference in their free energies:

$$G - G^0 = RT \ln \frac{C}{C_0}$$

If the standard concentration C_0 is 1 molar, the expression reduces to:

$$G = G^0 + RT \ln C$$

This expression can be applied to each of the substances involved in the reaction, so that $G_C = G_C^0 + RT \ln[C]$, $G_D = G_D^0 + RT \ln[D]$, etc., where $[C]$ denotes the concentration of component C, and so forth. (Note that the argument of the logarithm is dimensionless as it really expresses a ratio of two concentrations.) Equation (7.1) above then becomes:

$$\Delta G = \left(G_C^0 + RT \ln[C] + G_D^0 + RT \ln[D] \right)$$
$$- \left(G_A^0 + RT \ln[A] + G_B + RT \ln[B] \right)$$

(7.2)

Combining the logarithmic terms yields:

$$\Delta G = \left[\left(G_C^{\ 0} + G_D^{\ 0} \right) - \left(G_A^{\ 0} + G_B^{\ 0} \right) \right] + RT \ln \frac{[C][D]}{[A][B]}$$

Recognizing the first group of terms on the right-hand side as $\Delta G°$ reduces the expression to:

$$\Delta G = \Delta G^0 + RT \ln \frac{[C][D]}{[A][B]} \tag{7.3}$$

Letting the quotient of concentrations on the right-hand side be represented by Q:

$$\Delta G = \Delta G^0 + RT \ln Q \tag{7.4}$$

Equation (7.3) shows how the free energy of a chemical reaction varies with the concentrations of the reactants and products of the reaction. We have assumed that the reaction involves only one molecule of each of the various components. If two molecules of a component A, for example, enter the reaction instead of one, then A will appear twice in Equation (7.1) or (7.2), and, when the logarithmic terms for concentration are combined, we will have $[A]^2$ instead of $[A]$ in Equation (7.3). Each component in Equation (7.3) may have an exponent other than unity if the stoichiometry of the reaction requires it. For simplicity, we have assumed that the reaction involves only one molecule of each substance so that all concentration exponents are unity.

If all the reaction components are at standard, unit concentration, the quotient of concentrations, Q, itself becomes unity, and, since $\ln(1) = 0$, the rightmost term of Equation (7.3) vanishes, leaving $\Delta G = \Delta G°$ (as it should, by definition).

The quotient of concentrations, Q, will generally not remain constant during a chemical reaction. Unless the reactants A and B are constantly fed into the system and the end products C and D are continually removed, the concentrations of A and B will diminish as the reaction proceeds, and the concentrations of C and D will increase. When the reaction reaches its equilibrium state, two conditions are apparent: (1) the quotient of concentrations, Q, has become equal to the equilibrium coefficient, K_{eq}, and (2) the free energy change of the reaction, ΔG, has diminished to zero as the chemical gradient causing the reaction has been destroyed. At equilibrium:

$$Q = K_{eq}$$

$$\Delta G = 0$$

Equation (7.4) for the equilibrium state yields:

$$0 = \Delta G^{\circ} + RT \ln K_{eq}$$

$$\Delta G^0 = -RT \ln K_{eq} \tag{7.5}$$

Combining Equation (7.5) with Equation (7.4) gives:

$$\Delta G = -RT \ln K_{eq} + RT \ln Q$$

$$\Delta G = RT \ln \frac{Q}{K_{eq}} \tag{7.6}$$

Equation (7.6) says that the free energy change per mole of a chemical reaction is equal to RT times the logarithm of the ratio $\dfrac{Q}{K_{eq}}$. Since the free energy change per mole is the intensity factor for chemical energy, needing only to be multiplied by the capacity factor (the number of moles) to give the energy change, the quantity $RT \ln \dfrac{Q}{K_{eq}}$ is the chemical gradient itself, analogous to the height of a waterfall or the voltage of an electrical circuit. It is almost as if Q is the top of the chemical waterfall and K_{eq} is the bottom (though we are dealing with the logarithm of a ratio rather than a simple difference). The greater the difference between Q and K_{eq}, the greater is the free energy of the system. The free energy of a chemical reaction depends on the imbalance between the actual concentrations of reactants and products and the concentrations they would have at equilibrium.

Let us summarize from this perspective the course of a chemical reaction, noting how the progression of a chemical reaction tends to wear down its own gradient just as any other spontaneous change does. A chemical reaction, as represented in the following figure, is a progression in time, a change in chemical concentrations so that $Q \rightarrow K_{eq}$

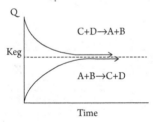

Time

If we start with a good deal of A and B, but no C and D, then the value of Q is initially zero, and, as A and B are converted to C and D, we progress from a zero value of Q toward a value of K_{eq}. If, on the other hand, we start with a great deal of C and D and very little A and B, the value of Q may be so high as to be greater than K_{eq}. In this case, Q will approach K_{eq} from above, as the reaction is reversed and C and D are converted to A and B. As the reaction progresses in either case (whether forward or reverse), the difference between Q and K_{eq} diminishes, as does the free energy obtainable from further reaction. At equilibrium, $Q = K_{eq}$ and $\Delta G = 0$, for $\Delta G = RT \ln \dfrac{Q}{K_{eq}} = RT \ln(1)$, and the logarithm of unity is zero. All chemical reactions tend toward an equilibrium state in which gradients have been reduced to zero, and there is no longer any chemical free energy available in the system.

The significance of the equation $\Delta G^0 = -RT \ln K_{eq}$ (Equation 7.5) is also worth pondering. It makes the powerful statement that if we know the standard free energy change of a chemical reaction, we also know the equilibrium constant or vice versa. The free energy available when all components are at standard concentration (which usually also means when all components are present at the same concentration) dictates the ratio of components at equilibrium. The free energy of a system determines, in chemistry as in everything else, the extent of possible change. The following table gives some examples at 25° C:

ΔG°	K_{eq}
−4092 calories/mole	1000
−2728	100
−1364	10
0	1
1364	0.1
2728	0.01
4092	0.001

A negative free energy change means that the reaction goes in the forward direction as written, for example A + B → C + D. The greater the free energy, the more complete the reaction will be; that is, the greater will be the concentrations of the end products relative to the reactants. If the free energy change is positive for the reaction A + B → C + D, then the change is equally negative for the reverse reaction C + D → A + B, and the equilibrium will have higher concentrations of A and B than of C and D. The equilibrium is a balance of chemical concentrations, and the free energy change tells us where the balance will be struck. It is to be noted that a rather modest negative free energy change determines a high preponderance of end products relative to reactants. For example, in the reaction between glucose and adenosine triphosphate (ATP) in the living cell, glucose-6-phosphate and adenosine diphosphate (ADP) are formed:

$$\text{Glucose} + \text{ATP} \rightarrow \text{Glucose} - 6 - \text{phosphate} + \text{ADP}$$

The standard free energy for this reaction is –3,700 calories/mole, which indicates that the product of the concentrations of glucose-6-phosphate and ADP will be about 517 times higher than the product of the glucose and ATP concentrations, if equilibrium is reached.

KINETICS AND THE EQUILIBRIUM CONSTANT

In a chemical reaction, the free energy determines the equilibrium constant, or, conversely, the equilibrium constant determines the free energy. The free energy does not determine, however, the rate at which the equilibrium is reached. To one accustomed to thinking about mechanical actions, this independence of free energy and rates of reaction may seem a little surprising. If a hillside gradient is steep, the brook flows quickly as well as with great energy. If the chemical gradient is steep, the reaction flows with great energy but may be rather slow. There is no necessary relationship between the energy of a chemical reaction and the speed at which it happens, except that both are affected in the same way by the concentrations of the reactants and products.

The effect of concentration on the rate of a reaction is some-times known as the "law of mass action." For two molecules to react, they must in some sense collide with one another, and the probability of such a collision is proportional to the concentration of each molecule. In the system:

$$A + B \underset{k_2}{\overset{k_1}{\rightleftharpoons}} C + D$$

the reaction is portrayed as revertible, so that just as A + B can form C + D, so C + D can form A + B. The rates of the forward and reverse reactions are given by:

$$v_1 = k_1 [A] [B]$$

$$v_2 = k_2 [C] [D]$$

where k_1 and k_2 are rate constants that are affected by temperature and other conditions of the system but are not affected by the concentrations of the components nor by time. It is noteworthy that k_1 and k_2 are not usually equal to one another, so that if all the components of the reaction have equal concentrations, the for-ward and reverse reaction velocities are generally not the same. Suppose that we start with just components A and B. Initially, then, there is no reverse reaction, and the rate of change in composition is given by v_1. But, as soon as some C and D have been formed, a reverse reaction begins, and the net rate of change in composition thereafter becomes $(v_1 - v_2)$. As the reaction proceeds, v_1 gradually gets smaller as the concentrations of A and B diminish, and v_2 gets larger as the concentrations of C and D increase. The net reaction velocity, $(v_1 - v_2)$, therefore decreases as the reaction proceeds. We have already seen that the free energy change of the reaction also decreases as the reaction proceeds. In this limited sense, because of their mutual dependence on the concentrations, the free energy change and rate of a reaction go hand in hand. The reaction proceeds until the net reaction, $(v_1 - v_2)$, becomes zero; that is, where $v_1 = v_2$, and a dynamic equilibrium, a balance of forward and reverse reactions persists, so that:

$$k_1[A][B] = k_2[C][D]$$

$$\frac{k_1}{k_2} = \frac{[C][D]}{[A][B]} = K_{eq}$$

Since k_1 and k_2 are both constants, their ratio is a constant, and since this constant is the quotient of equilibrium concentrations, it is the equilibrium constant, K_{eq}. Thus, the equilibrium constant is not only proportional (logarithmically) to the standard free energy change of the reaction, but also is proportional (linearly) to the ratio of the rate constants.

The rate of a chemical reaction can be changed dramatically by the presence of a *catalyst*, a substance that increases the rate of reaction without being used up by the reaction. The catalyst cannot change the free energy of the reaction because, if it could—by running the reaction one way with the catalyst and the other way without it—we might increase the free energy of the system without any alteration of the environment (and thereby invent perpetual motion). And since the catalyst cannot change the standard free energy, neither can it change the equilibrium constant or the ratio of the rate constants. Thus, a catalyst that increases the forward velocity constant of a reaction must increase the reverse velocity constant to the same extent.

The kinetic view of the chemical reaction provides an additional perspective on the dynamic nature of the equilibrium because the equilibrium is achieved by the forward and reverse reactions becoming equal to each other, not by their becoming equal to zero. At equilibrium, individual molecules of A and B still collide and still interact to form C and D, but individual molecules of C and D interact to form A and B at exactly the same rate. There is no net change at equilibrium, but individual molecules go on changing, changing their structure and position. We have seen how George Hevesy first demonstrated this in experiments with radioactive lead. In biochemistry, isotopic tracers show the dynamic nature of equilibria, such as that for the reaction between glucose and ATP:

$$\text{Glucose} + \text{ATP} \rightleftarrows \text{Glucose} - 6 - \text{phosphate} + \text{ADP}$$

After this system has reached equilibrium, if a small amount of radioactive C^{14} –labeled glucose-6-phosphate is added (which can be done without significantly altering the concentration of this component), the C^{14} label does not remain isolated in the glucose-6-phosphate but soon shows up in the glucose, too, indicating that individual molecules of glucose-6-phosphate are being converted to glucose even though the system as a whole is at equilibrium.

Whether we look at free energy changes or at rates of reaction, a chemical reaction is a revertible, balancing system that in principle can approach an equilibrium state from either direction.

It can be compared to a U-shaped tube filled in one arm with a liquid representing A and B, and in the other arm with a liquid representing C and D (see Figure 7.1):

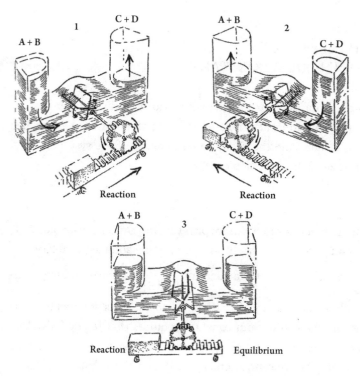

FIGURE 7.1. A hydraulic model of a chemical reaction. A chemical reaction A + B ⇌ C + D can flow in either direction depending on the relative concentrations of A and B on the one hand and C and D on the other. It is analogous to a reversible undershot waterwheel, which drives the ratchet to the right if levels of A and B are high (1) or to the left if levels of C and D are high (2). When the water levels balance (3), the system is at equilibrium and no work can be done, as the paddle wheel remains motionless. Reproduced from Evelyn L. Oginsky and Wayne W. Umbreit, *An Introduction to Bacterial Physiology*, p. 183. San Francisco: W. H. Freeman, 1954.

Whatever the starting levels of liquid in the arms (unless they happen to balance), there will be a movement of liquid one way or the other, with a consequent rotation of the paddle wheel in the middle and the potential for doing work until an equilibrium (balance) is reached.

THE EFFECT OF TEMPERATURE ON THE EQUILIBRIUM CONSTANT

The equilibrium constant gives the ratio of product concentrations to reactant concentrations when a reaction system has reached equilibrium (and the chemical free energy of the system has become zero). In a reaction of:

$$A + B \rightarrow C + D$$

the final equilibrium can involve varying concentrations of A, B, C, and D, but the concentrations must be adjusted in relation to one another, so that the quotient $\dfrac{[C][D]}{[A][B]}$ has the one value of the equilibrium constant, K_{eq}. If, for example, $[C]$ is doubled by some means, the equilibrium constant requires that $[D]$ be halved, or that $[A]$ or $[B]$ be doubled, or some combination of changes be made in the concentrations of A, B, and D so that the value of K_{eq} is maintained—if the system is to remain at equilibrium (an equilibrium that is clearly different from the one that we had before the concentration of C was doubled). That is the meaning of the equilibrium constant—that it retains a constant value for different equilibria involving varying concentrations of the substances involved.

For example, the equilibrium constant for the reaction $H_2O \rightarrow H^+ + OH^-$ is:

$$K_{eq} = \frac{[H^+][OH^-]}{[H_2O]} = 10^{-14}$$

when the concentration of water is, by convention, assigned a value of unity. For pure water, H^+ and OH^- each have a concentration of 10^{-7} molar (at a temperature close to $25°C$), so that the water has a pH (power of hydrogen) of 7. If enough HCl is added to the water to bring the H^+ concentration up to 10^{-6} molar (pH 6), the OH^- concentration has dropped to 10^{-8} molar (by a small reversal of the reaction), so that the product of the H^+ and OH^- concentrations remains at the equilibrium constant value of 10^{-14}.

The equilibrium constant, however, is not constant (usually) if the temperature is changed, and this is true for the dissociation of water. Most chemical reactions are either exothermic or endothermic, giving off or absorbing thermal energy, increasing or decreasing the temperature of the environment. By Le Chatelier's principle, the equilibrium system will respond to a temperature change by trying to reduce that change. If the temperature is increased, an exothermic reaction will tend to reverse somewhat to reduce its heat output, thus reducing the concentrations of end products and the value of the equilibrium constant. An endothermic reaction, on the other hand, will tend to go to greater completion if temperature is raised, thus raising the equilibrium constant. In qualitative terms, an increase in temperature increases the equilibrium constant for an endothermic reaction (a reaction in which ΔH is positive) and decreases it for an exothermic reaction (one in which ΔH is negative).

The quantitative effect of temperature on the equilibrium constant depends upon the ΔH value of the reaction. Temperature, equilibrium constant, and ΔH are all related to the free energy, $\Delta G°$, through the two relations:

$$\Delta G° = \Delta H° - T\Delta S°$$

$$\Delta G° = -RT \ln K_{eq}$$

(For simplicity, we will drop the superscripts and subscripts.) Equating the two right-hand sides of these equations yields:

$$\ln K = \frac{-\Delta H + T\Delta S}{RT} = -\frac{\Delta H}{RT} + \frac{\Delta S}{R}$$

To the extent that ΔH and ΔS remain constant over the temperature range considered, a plot of $\ln K$ against $1/T$ gives a straight line, with a slope of $-\Delta H/R$. Taking the derivative with respect to temperature of each side of the equation yields:

$$\frac{d(\ln K)}{dT} = \frac{\Delta H}{RT^2} \quad \text{(remembering that)} \quad \frac{d(x^{-1})}{dx} = -x^{-2}$$

Integrating this across a temperature range of T_1 to T_2 gives:

$$\ln \frac{K_2}{K_1} = -\frac{\Delta H}{R}\left(\frac{1}{T_2} - \frac{1}{T_1}\right)$$

If we wish, we can convert this logarithmic relation to the exponential form:

$$K_2 = K_1 e^{-\frac{\Delta H}{R}\left(\frac{1}{T_2} - \frac{1}{T_1}\right)}$$

In the preceding expressions, T_2 represents the higher temperature. The exponent is therefore positive if ΔH is positive (endothermic reaction) and negative if ΔH is negative (exothermic reaction). K_2 is therefore greater than K_1 for the endothermic reaction and less than K_1 for the exothermic reaction—as Le Chatelier's principle predicts.

The ionization of water itself reminds us of the pervasive influence of temperature on the equilibrium constants of chemical reactions. At a temperature a little below $25°C$, pure water has an ionization constant of 10^{-14}, so that the pH of pure water can be conveniently if rather loosely called 7 at "room temperature." However, the ionization coefficient for water varies strikingly with temperature:

Temperature $(°C)$	Ionization $K (\times 10^{14})$	pH of Pure Water
0	0.113	7.47
10	0.292	7.27
20	0.681	7.08
25	1.008	7.00
30	1.468	6.92
40	2.917	6.77
50	5.474	6.63

Using either the logarithmic or exponential equation that relates the equilibrium constant to temperature and ΔH, we can estimate the ionization coefficient for water at some other temperature, such as the normal human body temperature of $37°C$. From an equilibrium constant of approximately 1×10^{-14} at $25°C$, and assuming a ΔH of about 13,400 calories per mole for the heat of ionization (the ionization is an endothermic reaction), the equation predicts that the ionization constant for water at $37°C$ is 2.4×10^{-14}. This means that, at body temperature, the concentration of hydrogen ion in pure water is about 1.55×10^{-7}, which translates to a pH of 6.81. The blood plasma and other extracellular fluids of the body are not pure water, of course, but contain phosphate, carbonate, and protein buffers that maintain (with the long-term help of the kidneys) the pH at about 7.4.

The effect of temperature on the equilibrium constant is dramatically illustrated also in the equilibrium between liquid water and its vapor. The vapor pressure of water at $100°C$ is 760 mm Hg, which is why water (at sea level) boils at about that temperature. Given that the heat of vaporization of water at $100°C$ is about 540 calories per gram or 9,712 calories per mole, can we predict the vapor pressure of water at $20°C$? Yes, but not very accurately, because the ΔH for the vaporization increases somewhat as the temperature decreases from $100°$ to $20°C$. If we use the ΔH value at $60°C$ (midway in our temperature range), the approximation is quite good. With $\Delta H = 10,138$ calories per mole, the equation predicts that the vapor pressure of water at $20°C$ is 18.1 mm Hg. The actual value is 17.5 mm Hg.

FREE ENERGY IN ELECTROCHEMICAL REACTIONS

The free energy of chemical reactions is most directly evident in those reactions that can be arranged to generate a measurable electrical potential, as in a common flashlight battery. In an age of cell phones, laptop computers, and hybrid automobiles, no one doubts that the chemical reactions of batteries can do work. The amount of work they can do is equal to the free energy contained in their chemical reaction systems.

As Galvani and Volta discovered, dissimilar metals can give rise to electrical phenomena. When a copper (Cu) rod is placed in contact with a zinc (Zn) rod, the copper takes electrons from the zinc. The displacement of electrons from zinc to copper creates a voltage at their interface (copper negative, zinc positive) that opposes further transfer of electrons from zinc to copper, and the reaction is over almost as soon as it has begun.

If the copper rod is placed in a beaker containing a solution of zinc sulfate, essentially nothing happens:

$$Cu + Zn^{++} + SO_4^{--} \rightarrow \text{no reaction}$$

But if the zinc rod is placed in a beaker containing a solution of cupric sulfate, metallic copper begins to form on the zinc rod, and zinc ions begin to appear in solution:

$$Zn + Cu^{++} + SO_4^{--} \rightarrow Zn^{++} + Cu + SO_4^{--}$$

Electrons are passed from the Zn to the Cu^{++}, forming metallic Cu, while metallic Zn goes into solution as Zn^{++}. This illustrates again that copper will accept electrons from zinc more readily than zinc from copper, but, unlike the simple dry contact of zinc and copper rods, this reaction can go on for some time, involving a transfer of metal from solid to solution $\left(Zn \text{ to } Zn^{++}\right)$ or from solution to solid $\left(Cu^{++} \text{ to } Cu\right)$.

Suppose now that the zinc rod is mounted in a beaker containing a solution of $ZnSO_4$, and the copper rod is mounted in a separate beaker containing a solution of $CuSO_4$. A wire (preferably copper or zinc so as not to introduce a third metal into the system) connects the tops of the two rods. The copper rod pulls electrons from the zinc rod via the wire, and some of these electrons can be dumped onto Cu^{++} ions of the solution. The zinc rod can dump some of its resulting positive charge through Zn^{++} ions escaping from the rod to the solution. We have two "half-cell" electrochemical reactions beginning to combine into one reaction by passage of electrons through the wire from the zinc half-cell to the copper half-cell:

$$Zn \rightarrow Zn^{++} + 2e^-$$

$$Cu^{++} + 2e^- \rightarrow Cu$$

Adding these two half-cell reactions together gives the reaction for the full cell:

$$Zn + Cu^{++} \rightarrow Zn^{++} + Cu$$

However, this process no sooner begins but it stops because of the accumulation of negative charge in the copper sulfate and positive charge in the zinc sulfate, charges that resist further passage of electrons to the Cu^{++} or further escape of Zn^{++} ions into solution. To complete an electrochemical circuit, the two half-cells need to be connected internally by an ionic bridge as well as externally by the electrical wire, so that, as negative charge passes externally via electrons in the wire from zinc to copper, negative charge can pass internally via ions from the copper sulfate to the zinc sulfate (or positive ions pass from zinc sulfate to copper sulfate). An ionic bridge can consist of an inverted U-tube containing KCl solution held in an agar gel, so that positive and negative ions can pass through the agar bridge without any mass mixing of the contents of the two half-cells. With the ionic bridge, the electrical circuit is complete, and a flow of electrons can be sustained for some time by the chemical conversions of Zn to Zn^{++} and of Cu^{++} to Cu (see Figure 7.2).

Within the electrical cell, the zinc rod is the anode (positive electrode), and the copper rod is the cathode (negative electrode). But to the outside world of the electrical wire, the zinc rod is the negative source of electrons, and the copper rod is the positive terminal to which the electrons flow.

If a potentiometer is placed between the zinc and copper rods, the voltage generated by the reaction can be measured. A *potentiometer* is a special form of voltmeter. It contains a variable power source that opposes, through a galvanometer, the voltage generated by an unknown source. The voltage created by the potentiometer is adjusted until it just balances the unknown voltage, as shown by a zero deflection of the galvanometer. The potentiometer can be calibrated with known voltage sources. Its advantage over a simple voltmeter is that it reads the voltage of the unknown cell without drawing any current from it. Drawing current from a cell necessarily involves some decrease in its voltage due to internal resistance in the cell.

The voltage recorded from the zinc–copper oxidation-reduction cell is 1.10 volts, under standard conditions of $25°C$ and 1 molar solutions of $ZnSO_4$ and $CuSO_4$. The conversion of the chemical potential into an electrical potential gives us a clear and direct reading of the maximum work that the chemical reaction can do and hence of its free energy. For 1 mole of zinc oxidized to Zn^{++} (and 1 mole of Cu^{++} reduced to Cu), 2 "moles" of electrons must pass across a voltage of 1.10 volts. One mole of electrons carries the Faraday charge, $\mathcal{F}$, of electricity, which is 96,500 coulombs. The free energy (or maximum work, w_m) is the product

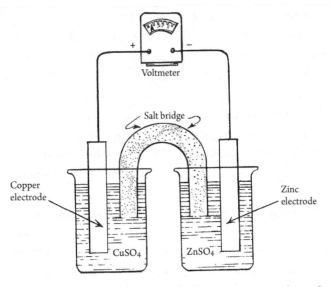

FIGURE 7.2. A zinc-copper electrochemical cell. The zinc-copper electrochemical cell produces an electrical potential and flow from a chemical oxidation-reduction reaction between zinc and copper. The copper has greater affinity for electrons than zinc, and the electron flow from the zinc plate to the copper plate through the external wire can be made to do work. Its potential can be measured by a voltmeter or potentiometer. Within the cell, the copper and zinc plates are immersed, respectively, in copper sulfate or zinc sulfate solution. As electrons are received via the wire by the copper plate, Cu^{++} from the copper sulfate combines with them and is deposited as metallic Cu on the plate. As electrons are lost to the wire by the zinc plate, metallic Zn becomes Zn^{++} and goes into solution as zinc sulfate. The salt bridge between the zinc sulfate and copper sulfate solutions completes the electrical circuit, permitting overall electrical neutrality to be maintained in the solutions as positive ions diffuse from right to left and negative ions from left to right within the bridge. Reproduced from Laurence E. Strong and Wilmer J. Stratton, *Chemical Energy*, p. 71. Reinhold, 1965.

of the charge (extensive factor) and electromotive force, $\mathcal{E}$ (the intensive factor or voltage difference):

$$w_m = -\Delta G° = n\mathcal{F}\mathcal{E}° = (2)(96{,}500 \text{ coulombs})(1.1 \text{ volts})$$
$$= 212{,}300 \text{ joules} = 50.7 \text{ kcal} / \text{mole Zn.}$$

Hence, the $\Delta G°$ for the reaction is –50.7 kcal/mole.

If the same oxidation-reduction of zinc and copper is carried out in a calorimeter, the heat of the reaction is $\Delta H° = -52.1 \text{ kcal} / \text{mole}$. The entropy change for the reaction is a small decrease, giving $T\Delta S° = -1.4 \text{ kcal} / \text{mole}$. As for all chemical reactions at constant temperature and pressure:

$$\Delta G° = \Delta H° - T\Delta S°$$
$$-50.7 = -52.1 - (-1.4) = -52.1 + 1.4 \text{ kcal} / \text{mole}$$

Thus, the free energy change measured directly by the electrical potential equals that calculated from thermochemical data. Whereas the free energy is deduced indirectly from the thermal measurements of enthalpy and entropy, it can be calculated (as before) directly from the electrical potential, without referring to enthalpy or entropy data. But it is instructive to compare the electrical and thermal results.

The electrical zinc–copper cell is reversible in two senses. The reaction could literally be reversed by a sufficient external electrical potential, so that electrons were made to flow from copper to zinc, and the cell reaction became:

$$Cu + Zn^{++} \rightarrow Cu^{++} + Zn$$

In such a case, the external electrical gradient would be doing work upon the chemical cell, building up its chemical gradient: that is, recharging the chemical cell. But the cell is reversible, too, in the sense that its voltage is being measured at zero current, so that no loss is incurred in the electrical potential or in the maximum work (or free energy) calculated from that potential.

A reversible reaction is one in which the loss of chemical gradient as the reaction proceeds is fully compensated by the building up of a free energy gradient outside the reaction itself. In such an ideal system, the total free energy of the world (system plus its environment) remains constant, and there is no increase in entropy and its associated bound (equipotential) energy. In the zinc-copper cell, the thermal data indicate that the cell itself loses a small amount of entropy. How is this possible? We must remember that the cell is in contact with its environment, both in the electrical wire that can do work on the environment and in the possible conduction of entropy (and associated heat) between cell and environment. The entropy of the cell has decreased slightly because entropy (and heat) has passed from the cell to the environment. The total entropy change is zero for the world of the cell plus its environment.

Of the zinc-copper cell's total energy (enthalpy) loss of 52.1 kcal/mole, 50.7 kcal/mole is work done on the environment, and 1.4 kcal/mole is heat conducted to the environment. Total free energy and total entropy remain constant in the reversible process. What the cell loses in free energy, the environment gains. What the cell loses in entropy, the environment gains. But suppose that the cell is short-circuited so that it can do no work on the environment, but entropy (and associated heat) is generated instead. Then the 50.7 kcal/mole of cell free energy is passed to the environment as bound equipotential energy. The change in entropy of the world (cell plus environment) is in this case:

$$\frac{50,700 \text{ cal/mole}}{298.15 \text{ deg}} = 170 \text{ entropy units}$$

In the short-circuited discharge of the cell, the world loses 50.7 kcal/mole of chemical free energy and gains 50.7 kcal/mole of bound thermal energy (associated with 170 e.u. of entropy).

Suppose now that two changes are made to the previous electrochemical cell. First, copper rods and copper sulfate solution are placed in both half-cells, so that the chemical constituents are identical in both. Then no chemical gradient exists between the two half-cells, there is no free energy, and no electrical potential. But now, second, let the concentration of copper sulfate be reduced in one of the half-cells to 0.1 molar while the copper sulfate is kept at 1.0 molar in the other half-cell. We have then a concentration gradient between the half-cells, and this concentration gradient has a potential free energy change of:

$$\Delta G = -RT \ln \frac{C_1}{C_2} = -(1.987)(298.15) \ln \left(\frac{1}{0.1} \right) = -1.363 \text{ kcal/mole}$$

Remarkably, this free energy difference can be made to do electrical work, for a potentiometer placed between the two copper posts shows a voltage difference of about 0.0296 volts (29.6 millivolts), the half-cell with the lower concentration of copper sulfate being negative. This can be calculated from the relation:

$$-\Delta G = n\mathcal{FE} \text{ or } \mathcal{E} = -\Delta G / n\mathcal{F}$$

Since the free energy change is proportional to the logarithm of the concentration ratio, squaring the ratio only doubles the free energy; that is, if a concentration ratio of 10 gives a free energy change of –1.363 kcal/mole (at $25°$C), then a ratio of 100 gives –2.726 kcal/mole. Every tenfold increase in the concentration ratio increases the free energy change by 1.363 kcal/mole. The free energy change is dependent only (at a given temperature) on the concentration ratio, not on the nature of the chemicals involved nor on the number of electrons exchanged per mole of chemical. The resulting electrical potential (voltage) is proportional directly to the free energy change but inversely proportional to the number, n, of electrons involved for electrical free energy (which in a fully compensated system equals the chemical free energy) is a product of its intensive factor (voltage) times its extensive factor (charge). If the number of electrons per ion is two instead of one, then the charge delivered per mole is doubled, and, for a given amount of electrochemical energy per mole, the voltage can be only half for a divalent ion as it is for a univalent ion.

It is certainly a fair question to ask why there is any voltage at all from a cell that contains no dissimilar metals. Why should one copper rod become negatively charged and the other positively charged? The answer can have nothing to do with the wire that connects the two copper posts in the outside world. It has everything to do with the different internal environments of the rods in the two half-cells. The one is bathed in a high concentration of Cu^{++} and SO_4^{--}, the other in a low concentration of those same ions. If we consider the half-cell reaction:

$$Cu \rightarrow Cu^{++} + 2e^-$$

the forward progress of this reaction is opposed by the presence of Cu^{++}, the more so in the half-cell that has the high concentration. Thus, the copper rod in the half-cell with the low concentration of Cu^{++} has a greater tendency to give up electrons than does the rod immersed in the high concentration, and electrons are pulled through the external wire from the low concentration side to the high. The electrical circuit is completed internally by the migration of positive ions from the anode and negative ions toward the anode.

Is there any heat of reaction (enthalpy change) for the electrical cell driven by a concentration difference? If there is, it should be small. The process does not involve a breaking of one kind of bond and the forming of another. It is an entropy-driven (thermally driven) process of scattering (equalizing concentrations), involving no new bond formation. If the electrochemical free energy is fully compensated by doing an equal amount of outside work (that is, if the process is reversible), then no overall free energy has been lost and no increase has occurred in the total amount of entropy in the world (system plus environment). The process is analogous to the reversible, isothermal expansion of a gas against an external pressure. The work done by the system on its environment (1.363 kcal/mole in the preceding example) would tend to cool the system except that the environment supplies the system with an equal amount of heat. The electrochemical system, like the gas, has lost free energy in doing work but gained bound (equipotential) energy in receiving entropy (heat). The environment has gained free energy and given up bound energy in passing entropy to the system. No entropy has been created in the reversible process, nor has any free energy been lost.

If the electrical flow of the concentration cell is short-circuited and prevented from doing any work on the environment, then the world (system plus environment) experiences a conversion of 1.363 kcal/mole of free energy to bound energy and an increase in entropy of:

$$\frac{1363 \text{ cal/mole}}{298.15 \text{ deg}} = 4.57 \text{ entropy units}$$

Chemical energy flows throughout the landscapes of our lives, uniting the wind with the waters and the earth with fire. It has been one of the great feats of human thought, drawing upon the genius of Lavoisier, Hess, Faraday, Carnot, Mayer, Helmholtz, Joule, Clausius, Gibbs, and many others, to see into the nature of this flow. These chemical rivers run ultimately into the sea of time, but in their passage through Earth give rise to the miracles of living organisms, and to that we shall turn in the concluding chapter.

Chapter 8

Biological Energy

Studies of living organisms played an important part in giving rise to the concept of energy, as shown in the work of Lavoisier and Laplace, Mayer, Helmholtz, and others. And the concept of energy in turn had a huge impact on the way that living organisms were subsequently viewed. All the myriad activities of life came to be seen within the context of a general natural energy, an energy that life shared with all of nature. Life did not make energy. Life is not the source of energy. Rather, it uses the gift of energy that is a part of the natural world in which it resides.

OPEN SYSTEMS AND THE PARADOX OF SUSTAINED BUT STABLE ACTIVITY

All living organisms are powered by chemical reactions. But all chemical reactions tend, like all forms of natural activity, to destroy their own gradients and achieve a condition of equilibrium, a condition of changeless stability. How, then, can life endure—life that is a sustained and more-or-less stable activity? All isolated chemical reactions and other phenomena tend to come to an end, but life goes on. It has continued through a course of 3 billion years or so. Why, since it is chemistry, has it not long ago followed the second law of thermodynamics to a condition of zero free energy devoid of change? How can life live with the second law, when the second law seems to say that life itself should cease?

To put it another way, life exhibits what seems to be a fundamental chemical paradox: it maintains activity and stability within the same system. In ordinary chemical systems, we have one or the other, either activity or stability, but not the two together. Either we have a chemical reaction going on (e.g. $A + B \rightarrow C + D$) with reactants being changed to products but with no stability of concentrations, or we have a chemical equilibrium (e.g. $A + B \rightleftharpoons C + D$) in which there is stability of concentrations but no longer any net change since the forward and reverse reactions balance each other. In isolated chemical systems, the stability of equilibrium and the changing concentrations of reaction are mutually exclusive. Equilibrium implies no further conversion of free energy to bound, while reaction involves a decrease in free energy. When one considers the stability of living organisms, one may be tempted to describe them as being in equilibrium. But their activity and continual use of free energy belies this. Life is a nonequilibrium, for activity and free energy utilization persist. Yet it is steady (neglecting long-term developmental or evolutionary changes). How do we see this in chemical terms?

The answer to the paradox of combined activity and stability lies in one of the most basic of life's many secrets: life is an open system, with a continual interaction and exchange between the organism and its environment. An ordinary chemical system is usually a closed system, a system that is self-contained (like the materials inside a test tube), isolated from material exchange with its environment. But a living organism is open to its environment and exchanging materials with it, taking in some materials and giving off others. There is in fact no such thing as an isolated living system. We refer to living organisms as if they were isolated beings—you or me or my cat Magellan—but we are none of us bounded by our outward form, nor are any of our cells closed off by their plasma membranes. Life is not an isolated state of being, but a process of interaction, a communion between organism and environment.

Whereas a closed system, being complete in itself and isolated from any external sources of free energy, must tend to go to a condition of equilibrium in which all its gradients have been destroyed, an open system, if it receives constant inputs of free energy from gradients outside itself, may go to a "steady-state" instead of an equilibrium. What is a steady-state? It is a condition in which gradients become constant but not zero, where flow continues but at a constant rate throughout the system, and where the concentrations of components remain steady because each component is produced at the same rate as it is broken down. The differences between the closed and open systems are represented schematically in the figure below.

	Closed system	Open system
Initial condition	A→B→C→D→E	A ⊢→B→C→D⊣→E
Final condition	A⇌B⇌C⇌D⇌E	A⇌B⇌C⇌D⇌E
	equilibrium gradients: zero no net flux	steady-state gradients: steady constant net flux

In each of the systems is a sequence of reactions, A producing B, B producing C, and so on, with E as an ultimate end product. Initially, the reactions in both systems are all in the forward direction, assuming that the initial concentration of A is much greater than that of the subsequent components. But as B, C, D, and E are produced from A, back reactions begin to build up. In the closed system, these back reactions increase until an equilibrium is achieved. The concentrations of A, B, C, D, and E remain constant thereafter because there is no further net flux between any of the components. In the open system, however, the back reactions do not build up so far because—and this is a large because—the source, A, and the ultimate product or sink, E, lie outside the rest of the system and are held constant by factors in the environment of the open system itself. In other words, as A is converted to B, something in the environment prevents the concentration of A from falling, and, as D is converted to E, something prevents the concentration of E from rising. Under such circumstances, the system can achieve a steady-state in which the concentrations of B, C, and D remain constant because there is a constant net flux through the system—because, for example, C is produced from B and simultaneously broken down to D at rates that are constant and equal to one another.

An open, steady-state system is like a river that is always flowing but that retains a constancy of its basic characteristics. The Greek philosopher Heracleitus asserted that we never step twice into the same stream, for the stream's water is constantly flowing in and out, and the water of one moment is not that of the next. But the level and form of the stream remain constant (excluding erosion or evolution) so long as inflow and outflow remain the same. And so it is with living things: their materials are constantly flowing in and out (the water in the human body turns over about every 3 weeks), yet the levels of materials are maintained nearly constant, and the patterns or architecture of the living stream survive the continual turnover of its molecular pieces.

A living cell or a living organism is an open chemical system that achieves a nearly steady-state condition. It is enabled thereby to live not only with the second law of thermodynamics, but also, by its very direction, driven by the gradients of chemical reactions but achieving in its open communion with its environment a synthesis of activity with stability. The ultimate gradient is in the environment, between the source, A, and the sink, E, and the living organism is absolutely dependent upon the maintenance of that environmental gradient. Life is not organism alone, but organism and environment. The thing that lives is a pattern of components, B, C, and D, which constitute a channel between A and E. Life in this material sense is neither the beginning nor the end, but that which goes between, a beautiful, swirling eddy sustained by the energy gradients of the world. "No man is an island, entire in itself." Neither is any living thing. Life is lived by openness. It is chemistry, but chemistry intimately in touch with its world.

How can the external gradients that drive the open, living state remain constant in spite of the flow of reactions? How can A and E remain constant when the very reactions flowing between them should diminish A and increase E? For life viewed as a whole on our Earth, the Sun is almost the entire ultimate source, A, of the driving energy gradient, while the ultimate sink, E, is outer space. The Earth and the life upon it reside between Sun and space, receiving light and giving off heat (radiant light of longer wavelength). We know that the Sun is not completely constant, and we believe that it has had an origin and will someday have an end. But it requires billions of years for a star like the Sun to run its course from beginning to end, so that it is nearly constant on lesser scales of time. Its output of energy has probably remained almost constant during the 3 billion years that life has existed on the Earth. On the other end of the gradient, the immensity of outer space assures that it remains a constant sink for the light and heat flowing out from our solar system.

The entire world of life is an open system residing within the nearly constant energy gradient maintained by the Sun and outer space. But within this primary gradient that drives the green world of chlorophyll there develop a host of subsidiary gradients that power the lives of individual organisms. The green plants face the Sun's gradient directly, while others, including the animals, fungi, and many microorganisms, find themselves living, energetically speaking, in the shade. In the extraordinary web of life, one organism complements another: what is end product, E, for one living open system is starting material, A, for another. The result of this overlapping of interests is that materials are never used up, are never converted entirely to one form or to some final equilibrium mixture, but are passed on cyclically from one organism to another. We have what is commonly called the *balance of nature*. As the grandest example of all, the open systems of animals and green plants reside within a common cycle and complement one another, as depicted in the figure below.

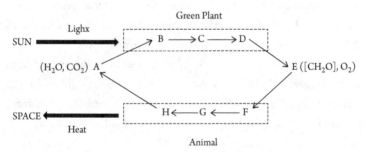

The green plant takes in water and carbon dioxide and, capturing the energy of the Sun's light, synthesizes (with a few minerals) its various plant materials, giving off oxygen and ultimately carbohydrates and other nutrients as gifts to the animal world. The animal takes in the oxygen and carbohydrates (the figure omits the proteins, fats, and other nutrients) and converts them to water and carbon dioxide, using the energy of oxidation to drive its own chemical syntheses, mechanical activities, and production of heat. The entire coupled system of plants and animals is an open system dependent upon the gradient from Sun to space. Within the overall system each individual organism, and each cell within an individual, is an open system driven by outside energy gradients but passing material components cyclically from one to another.

One may choose any element that occurs in biological organisms and describe how it is passed from one organism to another. One has then an oxygen cycle, a carbon cycle, a nitrogen cycle, and so forth. These cycles ultimately are very complex because they involve all of the several million different species of organisms, and they are all to some degree interrelated as different elements are passed together from one organism to another.

Wherever one looks at the activities of the living world, one is seeing open systems with their accompanying flows of energy and materials, be it in single cells, individual plants or animals, or communities of organisms such as a pond, a forest, or a prairie. Ultimately, the materials are recycled, but the energy flows from Sun to outer space. If we try to add up the entire chemistry of living nature, we are confronted by an astonishing fact. For each activity of any living organism, some kind of chemical equation could be written. For example, for photosynthesis we have:

$$\text{Light}$$
$$\downarrow$$
$$6\,CO_2 + 6\,H_2O \rightarrow C_6H_{12}O_6 + 6\,O_2$$

and for respiration we have the reverse:

$$C_6H_{12}O_6 + 6\,O_2 \rightarrow 6\,CO_2 + 6\,H_2O$$
$$\downarrow$$

Heat

But if we add these two equations together, they cancel each other out entirely, except for the Sun's light, which has been re-radiated to outer space as heat. In nature, green plant (and microbe) photosynthesis and plant and animal (and fungal and microbe) respiration are so balanced that the one restores the other. And what is true for photosynthesis and respiration is true also for all the reactions of the interlocking open systems of living nature—with exception given only to long-term interfaces between biological and geological processes involving the formation of coal, oil, carbonate, or silicate deposits. To a very close approximation, if we could write a chemical equation for each activity of every living organism on Earth and were able to sum up the whole scheme, the result would be zero, for everything produced in one equation would be consumed in some other. In terms of material change, life all adds up to nothing. Does that mean that life is "the sound and the fury signifying nothing?" Or does it mean that it is the grandest example of how much can be had for so little?

FREE ENERGY UTILIZATION IN THE OPEN STEADY-STATE

Mechanical work done by a living organism is achieved by the expenditure of chemical free energy. A person engaged in heavy physical activity may expend 4,000 kcal/day or more. But even a person asleep in bed requires considerable free energy just to maintain life—the resting organism still breathes oxygen, oxidizes food, and produces heat. This resting metabolism of a living thing can be viewed as the free energy flow required to maintain an open system in a steady-state removed from a point of equilibrium. The stability of an equilibrium is achieved by the elimination of gradients, with free energy expenditure necessarily reduced to zero. But the open system achieves stability in a steady-state, maintaining a continuous flow of reactions across a gradient of chemical concentrations. The amount of energy needed to maintain this steady-state depends upon the height of the gradient (the extent to which chemical concentrations are kept removed from their equilibrium levels) and the amount of reaction that is allowed to flow across that gradient.

Suppose that a living organism is a network of reactions in steady-state. One reaction within that system involves the conversion of A and B to C and D, but this reaction is not allowed to reach equilibrium because of the continual influx of A and B from other reactions and the continual efflux of C and D to other reactions. This single reaction is hence kept in steady-state, with a concentration quotient, Q, that is different from the equilibrium constant, K_{eq}. The free energy change (per mole) for this reaction is then:

$$\Delta G = RT\ln\frac{Q}{K_{eq}}\,\text{calories/mole}$$

The rate at which free energy is expended by this reaction is obtained by multiplying this free energy change per mole by the rate, ρ, of the reaction expressed in moles per minute:

$$\text{Rate of energy expenditure} = \rho RT\ln\frac{Q}{K_{eq}}\,\text{calories/minute}$$

We have calculated the "resting metabolism" for a single reaction in steady-state. If we could sum up all the reactions going on in the open system of the living organism, we would have determined the resting energy expenditure of that organism. That is, of course, impossible, but we can see by this argument the basis for why a living organism requires energy even when it is at rest.

COUPLED REACTIONS AND THE PARADOX
OF INCREASING ORDER

The first challenge to life presented by the second law of thermodynamics is how to maintain continual and sustained activity in the face of the tendency of gradients to diminish. The living organism solves that problem by being an open system channeling outside energy gradients and by living in a steady-state (more or less) rather than settling into the inactivity of equilibrium. Life is a living stream.

But, as if to double the paradox, the streambed of life seems to lie above the energy levels of much of its environment, like a river flowing above its own landscape (such streams actually exist in the drained and shrunken fenland of eastern England). Life requires an orderly arrangement of complex materials, an architecture that is very different from its nonliving environment, but the second law seems to say that differences should tend, with activity, to disappear, leveling the landscape of nature. How can the living organism maintain its vital differences, its architectural integrity, when the second law says that activity should disperse these differences between the organism and its environment? The following figure illustrates the question. How is it possible for life to be not only an open cascade of reactions from source to sink, but also a cascade with complex components, X, Y, Z, lying above the energy levels of the stream itself?

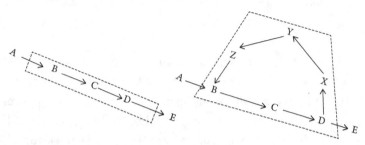

The answer lies in finding a way to couple one free energy gradient to another. Since the quantity of free energy depends upon the product of a gradient and the amount of charge flowing across it, it is possible for a large chemical stream flowing across a small gradient to drive a small chemical stream up a larger gradient—if means can be found to couple the one stream to the other. As we have seen before, an electrical step-up transformer can produce a large voltage from a small one, provided the current flowing across the high-voltage difference is correspondingly smaller than the current across the low-voltage difference. So it is with the chemical reactions in living organisms.

Let us see some examples of how this works. The living cell requires for its architecture the synthesis of large molecules such as proteins, nucleic acids, and polysaccharides. But these macromolecules reside at higher energy levels than the smaller molecules from which they are made—that is, it requires the expenditure of free energy to make the large molecules from the small, to synthesize (for example) a polysaccharide such as the glycogen of our muscles and liver from the small precursor glucose molecule. If glucose is released from a polysaccharide (containing

n units of potential glucose) by hydrolysis with water, the free energy change is –4,500 calories per mole of glucose:

$$(\text{polysaccharide})_n + \text{water} \rightarrow (\text{polysaccharide})_{n-1} + \text{glucose}$$
$$\Delta G^{\circ} = -4500 \text{ cal / mole}$$

Under standard conditions, this is a spontaneous (down-gradient) reaction with release of free energy. But then the reverse of this reaction is up-gradient, requiring 4,500 calories for each mole of glucose incorporated into polysaccharide. In this sense, the polysaccharide is at higher energy than its glucose constituents, and the cell must have energy available from some other source in order to make the large molecule from the small. But how can the other energy source be coupled to the desired synthesis of the macromolecule?

Suppose that we have the following reactions:

$$A + B \rightarrow C + D \quad \Delta G^{\circ} = + 5000 \text{ calories / mole}$$
$$E + F \rightarrow G + H \quad \Delta G^{\circ} = - !0,000 \text{ calories / mole}$$

If there were some way to add the two reactions together, then their free energy changes would be additive, too, giving a reaction that would flow downhill with a free energy change of –5,000 calories/mole in spite of the production within the system of C and D from A and B.

The living cell has extraordinary means for combining reactions together so as to utilize the free energy of one to drive another. For example, an aerobic cell (one that utilizes O_2) has at its disposal the powerful downhill reaction of oxidizing substrates such as glucose:

$$C_6H_{12}O_6 + 6\,O_2 \rightarrow 6\,CO_2 + 6\,H_2O \quad \Delta G^{\circ} = -686,000 \text{ cal / mole}$$

If there were some way to add this reaction to that converting glucose to polysaccharide, the cell would have a downhill system in which part of the available glucose was oxidized while part was used for building polysaccharide. In fact, the oxidative reaction is so powerful that only a small part of the glucose would have to be oxidized in order to convert the rest to polysaccharide.

But if reactions are to be added together, they must be chemically coupled in some material way. As a motor car must have a transmission, with gears and rods that link its driving engine to its wheels, so the cell must have a chemical linkage that couples one reaction to another. This linkage can take the form either of a chemical substance that is shared between the two reactions (a common intermediary component), or of a trans-membrane concentration gradient shared by the two reactions. We will consider first examples of the common intermediary component.

In being coupled together, biochemical reactions necessarily lose their isolated character and become part of a more complex system. The individual reactions are often changed somewhat by the introduction of the coupling component. Among the most common coupling agents in the chemistry of the living cell are molecules that transfer phosphate or hydrogen from one substrate to another. Here, we will consider the transfer of phosphate groups that can link the two processes of the oxidation of glucose and the synthesis of glycogen polysaccharide from glucose.

Phosphate transfer is accomplished most commonly by the agency of adenosine triphosphate (ATP) and its partner, adenosine diphosphate (ADP). ATP has a strong propensity to lose phosphate groups by the reaction:

$$\text{ATP} + \text{water} \rightarrow \text{ADP} + \text{phosphate} \quad \Delta G^\circ = -7000 \text{ cal} / \text{mole}$$

It is common to say that the ATP has a higher free energy than the ADP and to refer to the terminal phosphate group of ATP (and also that of ADP) as being held by a high-energy phosphate bond. Since the conversion of ATP to ADP releases (under standard conditions) 7,000 cal/mole of free energy, the restoration of ATP from ADP requires an equal amount.

By a complex sequence of reactions, the cell can couple the oxidation of nutrients such as glucose to the formation of the higher energy ATP from the lower energy ADP. The oxidation of glucose has a $\Delta G°$ of −686,000 cal/mole, while the formation of ATP from ADP requires only 7,000 cal/mole, hence the cell is able, in terms of energy, to form many moles of ATP (38 it is thought) from each mole of glucose that is oxidized. Since 38 moles of ATP require 266,000 calories for their formation, the oxidation of glucose coupled to this amount of ATP regeneration would still have a powerful downhill gradient of $686,000 − 266,000 = 420,000$ cal / mole.

As depicted below, the ATP–ADP transformations form a coupling device between the downhill oxidation of glucose and the uphill formation of polysaccharide from glucose. The downhill oxidation of glucose pushes ADP uphill to ATP, and the downhill conversion of ATP to ADP pushes glucose uphill to polysaccharide.

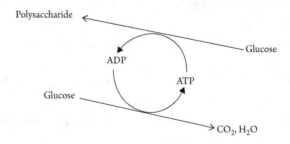

We will consider just one of the reactions linking the oxidation of glucose to ATP formation. This reaction is the oxidation of 3-phosphoglyceraldehye (PGAL) to 3-phosphoglyceric acid (PGA), both of which are compounds containing three carbon atoms produced from a prior splitting of the 6-carbon glucose. We will represent the compounds simply as a radicle, R, attached to an aldehyde (CHO) or acid (COOH) group. The oxidation of the aldehyde to the acid proceeds by the addition of water and the removal of hydrogen (which is passed to the hydrogen-receptive coenzyme, nicotine adenine dinucleotide [NAD]):

$$
\begin{array}{cc}
O & O \\
\| & \| \\
R-\ C\ -H\ +H_2O\ \rightarrow\ R-\ C\ -OH\ +2H & \Delta G^\circ = -7000\ cal\,/\,mole
\end{array}
$$

The reaction written is the one that would occur if the oxidation were not coupled to the formation of ATP; it involves a substantial drop in free energy. But, in the cell, the reaction is normally modified to introduce phosphate instead of water:

$$
\begin{array}{ccc}
O & O & O \\
\| & \| & \| \\
R\ -C\ -H\ +H_3PO_4\ \rightarrow\ R\ -C\ -O & -P\ -OH\ +2H \\
 & & | \\
 & & OH
\end{array}
$$

The phosphate so introduced is of high energy and is immediately transferred to ADP:

$$
\begin{array}{cccc}
O & O & & O \\
\| & \| & & \| \\
R\ -C\ -O\ -P\ -OH\ + ADP\ \rightarrow\ R\ -C\ -OH\ + ATP \\
 & | \\
 & OH
\end{array}
$$

The result is that the aldehyde is oxidized to the acid as before, but the process is modified to incorporate the conversion of ADP to ATP. Each of the two steps in the modified process has a negative but very small free energy change; each step is therefore spontaneous, but most of the free energy that would otherwise have been released as heat (about 7,000 cal/mole) has been retained as the chemical energy of ATP. The gradient of the aldehyde oxidation has been compensated by the building of a high-energy phosphate gradient, so that most of the free energy is retained but shifted from one chemical gradient to another.

To complete the linkage of the downhill glucose oxidation to the uphill synthesis of polysaccharide via the ADP–ATP coupling mechanism, we now consider how ATP drives the synthesis of glycogen from glucose. This process involves four steps. The first step is a direct reaction between ATP and glucose:

$$ATP + glucose \rightarrow glucose - 6 - phosphate + ADP$$
$$\Delta G = -3700 \text{ cal / mole}$$

This reaction activates the glucose by raising its energy level to that of glucose-6-phosphate, but it reduces the energy level of ATP to a greater extent, so that the overall reaction is spontaneous. The reaction could be regarded as the sum of the following two reactions added together:

$$glucose + phosphate \rightarrow glucose - 6 - phosphate + water \quad \Delta G^{\circ} = +3300 \text{ cal / mole}$$

$$\underline{ATP + water \rightarrow ADP + phosphate \qquad\qquad\qquad \Delta G^{\circ} = -7000 \text{ cal / mole}}$$

$$glucose + ATP \rightarrow glucose - 6 - phosphate + ADP \qquad \Delta G^{\circ} = -3700 \text{ cal / mole}$$

The coupling occurs by replacing the two hypothetical reactions (the first of which cannot occur by itself under standard conditions) with a reaction which represents their sum, but which is nevertheless a new reaction—the glucose now reacting with the phosphate of ATP rather than with free phosphate, and the ATP now reacting with glucose rather than with water. In this manner, an ATP generated previously by the oxidation of one glucose is used to activate a second glucose to the higher energy of glucose-6-phosphate.

The glucose-6-phosphate is then converted in a second step to a similar compound, glucose-1-phosphate, in which the phosphate group is located on the first carbon of the molecule instead of on the sixth:

$$\text{glucose} - 6 - \text{phosphate} \rightarrow \text{glucose} - 1 - \text{phosphate}$$
$$\Delta G^{\circ} = +1700 \, \text{cal} / \text{mole}$$

This reaction appears to require coupling to some other energy source for it has a small positive free energy change, but this is true only under standard conditions, that is, when the reactant and product are present in equal concentrations. The K_{eq} for this reaction is about $\frac{1}{18}$ (a 10-fold difference in concentration has a free energy equivalence of 1,364 cal/mole), which means that equilibrium exists when there are 18 parts of glucose-6-phosphate for every 1 part of glucose-1-phosphate. If, however, there are more than 18 parts of glucose-6-phosphate for every 1 part of glucose-1-phosphate, the reaction will proceed toward the production of glucose-1-phosphate. Here is a case where the standard free energy of a reaction is small enough that the necessary gradient can be created by keeping the concentration of the reactant modestly higher than that of the end product. This can be accomplished for this reaction if glucose-6-phosphate is rapidly produced from glucose while glucose-1-phosphate is quickly drained by further reactions.

The glucose-1-phosphate is in fact rapidly drained by a third step, a further activation sending it on its way toward glycogen or some other polysaccharide. The glucose-1-phosphate reacts with uridine triphosphate (UTP), which is a nucleotide similar to ATP, to form a uridine diphosphate–glucose (UDP-glucose) complex:

$$UTP + glucose - 1 - phosphate \rightarrow UDP - glucose + diphosphate$$
$$\Delta G° = 0 \text{ cal / mole}$$

Although this reaction has little or no free energy gradient at standard concentrations of reactants and products, it can be made to proceed in the forward direction by keeping the concentration of one of the products low. This is done by a further reaction that splits the diphosphate into two monophosphate ions, a reaction that involves a change in free energy of –5,000 cal/mole. This second reaction pulls the first one along by keeping the concentration of diphosphate low.

The UDP-glucose represents the highest energy level to which glucose is activated. In a fourth and final step, this high-energy form of glucose condenses with a growing glycogen chain, building the glucose into the glycogen and releasing free uridine diphosphate:

$$UDP - glucose + \left(glycogen\right)_n \rightarrow \left(glycogen\right)_{n+1} + UDP$$
$$\Delta G° = - 2500 \text{ cal / mole}$$

where n and $n+1$ represent the number of glucose residues in the glycogen.

The four-step process of raising glucose to the higher energy level of glycogen is summarized in the following figure.

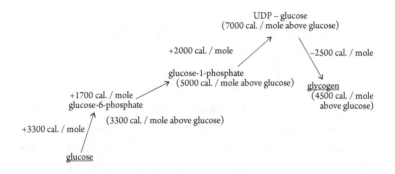

The glycogen lies at an energy level about 4,500 cal/mole above that of the original glucose, but it is produced by reactions each one of which is downhill at the concentrations employed. These reactions are:

		$\Delta G°$
1.	glucose + ATP → glucose − 6 − phosphate + ADP	−3700 cal/mole
2.	glucose − 6 − phosphate → glucose − 1 − phosphate	+1700
3a.	glucose − 1 − phosphate + UTP → UDP − glucose + diphosphate	0
3b.	diphosphate → 2 monophosphate	−5000
4a.	UDP − glucose + $(glycogen)_n$ → $(glycogen)_{n+1}$ + UDP	−2500
4b.	UDP + ATP → UTP + ADP	0

The last reaction is a subsidiary one that is necessary if UDP is to be recycled to UTP for reaction with further glucose-1-phosphate.

The reaction is accomplished at the expense of ATP, which gives up its terminal phosphate to UDP. If we sum up all six reactions, the net result is:

$$\text{glucose} + (\text{glycogen})_n + 2\text{ATP} \rightarrow (\text{glycogen})_{n+1} + 2\text{ADP} \\ + 2 \text{ monophosphate}$$

The total free energy change is −9,500 cal/mole. Everything cancels out of the final reckoning, except that one glucose is converted to glycogen at the expense of two ATP being converted to ADP. Two moles of ATP converted to ADP by hydrolysis with water represent 14,000 calories of free energy. Of this 14,000 calories, the formation of glycogen is able to store 4,500 calories as chemical work done, with the release of the remaining 9,500 calories as heat. In such arithmetic, we see the basis for the living organism always having to produce heat, whether or not it is involved in any apparent mechanical activity.

This example of the way that glucose oxidation can be coupled to the formation of highly ordered macromolecules in the living cell illustrates how life uses a chemical free energy gradient to create structures which are more complex than the materials in their environment. It is not our purpose to explore the many hundreds of reactions that go on in the living cell, but only to see the energetic principles involved. A further point, however, should be made. Each of the reactions above requires a specific enzyme, itself a highly complex protein molecule, to catalyze the reaction, allowing the reaction to proceed at a rate that is many times (often many billions of times) faster than it would go without the catalyst. The enzymes specify what actually happens among all possible reactions that might occur. For example, the reaction of ATP with water has a free energy gradient of −7,000 cal/mole, while the reaction of ATP with glucose has a gradient of only –3,700 cal/ mole. Why would any ATP react with glucose when it could, with greater gradient, react with the water that is all around it? The answer lies in the enzyme that specifically brings ATP together with glucose, greatly speeding this reaction over the more energetic one with water. Furthermore, many enzymes are sensitive to molecular signals (as from end products of a reaction sequence) that change their rates of reaction to accord with the needs of the living cell. Such homeostatic feedback signals allow the cell to maintain the necessary conditions for its life.

The coupling mechanisms that we have noted in the formation of glycogen could be classified into two types, which we might call *summation coupling* and *sequential coupling*. In summation coupling, two reactions that involve common components are replaced by a new reaction that is the sum of the previous two:

$$A + X \rightarrow C + Y \quad \Delta G^{\circ} > 0$$
$$\underline{B + Y \rightarrow D + X \quad \Delta G^{\circ} \ll 0}$$
$$A + B \rightarrow C + D \quad \Delta G^{\circ} < 0$$

Although the first reaction has a positive free energy, the second has a greater negative free energy, and their sum consequently is a spontaneous reaction. An example of summation coupling was seen in the formation of glucose-6-phosphate from glucose (reaction 1).

In sequential coupling, an end product of a first reaction is a substrate for a second reaction, so that if the first reaction has a positive free energy change, it can nevertheless be pulled along by a second reaction having a greater negative free energy change. By keeping the concentration of the end product of the first reaction low, the second reaction allows the first to operate spontaneously (at the nonstandard concentrations):

$$A + B \rightarrow C + Y \quad \Delta G° > 0$$

$$\underline{Y \rightarrow D \qquad \Delta G° \ll 0}$$

$$A + B \rightarrow C + D \quad \Delta G° < 0$$

An example of sequential coupling was seen in the formation of UDP-glucose from glucose-1-phosphate (reactions 3a and 3b).

In summary, the creation of high-energy order in an open chemical gradient might be depicted as a kind of chemical waterwheel pumping selected materials uphill, as a physical water wheel might drive a chain of buckets carrying water to a higher reservoir (Figure 8.1).

Polysaccharides, proteins, nucleic acids, etc.

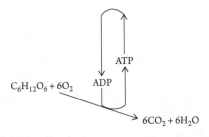

$$C_6H_{12}O_6 + 6O_2$$

ATP

ADP

$$6CO_2 + 6H_2O$$

It was thought by many eminent scientists, even into the early years of the 20th century, that life must have some vital force that allows it to generate itself from the lifeless materials of its environment. But life does not have to defy nature, for nature is nurturing of life. As Lawrence J. Henderson pointed out in his classic study, *The Fitness of the Environment*, nature is not something that life has

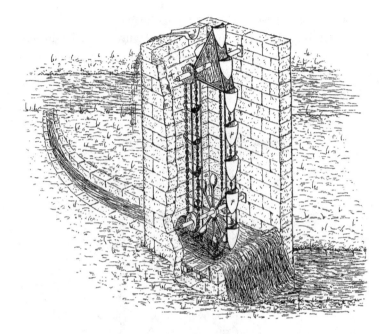

FIGURE 8.1. An ancient undershot waterwheel pump. This ancient machine illustrates the principle that a large volume of water flowing across a small gradient can lift a small volume of water to a much higher level. It also can serve as a metaphor for the metabolic energetics of a living organism, which couple chemical flows in such a way as to lift molecular compounds (e.g., proteins and nucleic acids) to high levels of energy and organization. Reproduced from Abbott Payson Usher, *A History of Mechanical Inventions*, p. 154. Beacon Press, 1959.

to overcome, but is uniquely suited, through the characteristics of many of its most common substances, such as water and phosphate and carbonate, to provide a matrix in which the activity, stability, and organization of living things can occur. Likewise, the laws of energy and entropy are also the home for life.

Still, the more one learns of life, the more one stands in awe of the seeming improbability of it all. The essence of this feeling has been caught in a cartoon, a version of which appeared long ago in the *Saturday Review of Literature* (see Figure 8.2). The drawing depicts sand dropping from the top to the bottom of an hourglass, flowing as all things do in time. Yet, behold, the falling sand is building castles:

FIGURE 8.2. An hour-glass cartoon. We do not expect sand falling freely across a gravitational gradient to create castles, but material falling across organized chemical gradients can create the structural intricacies of living things. Reproduced from a free Internet image and similar to one appearing many years ago in *The Saturday Review of Literature*.

VECTORIAL CHEMISTRY OF MEMBRANES: COUPLING REACTIONS WITH TRANSPORT

In addition to coupling reaction gradients by sharing intermediate components, as we saw in summation and sequential coupling, the living cell can couple gradients of chemical reaction to those of physical transport by means of the architectural design of its cell membranes.

As an open system, the cell must continually exchange materials with its environment but at the same time maintain concentrations of materials within itself that are very different from those around it. The interior of the cell not only contains the highly ordered nucleic acids, proteins, and other macromolecules that it has synthesized at considerable energy cost, but also a unique selection of small molecules and ions. The animal cell, for example, contains a high concentration of potassium (K^+) relative to its environment, but a low concentration of sodium (Na^+). The living cell must guard its own identity from the tendency of all materials to spread by diffusion down concentration gradients. It does this by maintaining around itself a lipid membrane into which are embedded proteins that serve as enzyme catalysts and transport channels. Across these semipermeable membranes, some materials can readily diffuse, while others cannot. Further membranes bound organelles within the cell, permitting different concentrations and activities to be maintained in different parts of the cell.

The membranes of cells make possible not only the compartmentalization of materials and activities, but also the alignment and orientation of enzyme catalysts. If an enzyme is free in solution, it can rotate in any direction as the result of collisions with other molecules, substrates and enzyme can approach one another from any direction, and end products can leave the enzyme in any direction. This is what we generally expect in chemical reactions: that there is no special direction in space to the reaction. But when enzymes are aligned in a membrane structure with active sites of the enzymes facing specifically toward the inside or outside of the membrane, a reaction can take on a specific direction across the membrane. The chemistry then becomes vectorial in nature: that is, it has a direction as well as a rate. This makes possible some of the most surprising processes of the living cell, in which chemical reactions are coupled to physical transmembrane transport, the membrane itself acting as the coupling mechanism between reaction and transport.

As an example, suppose an enzyme for catalyzing the reaction between ATP and water to form ADP and phosphate is built into the structure of a membrane. The enzyme, called an ATPase, is constrained by the membrane so that ATP is received and ADP and phosphate are given off specifically on one side (call it the inner side) of the membrane. Water enters the reaction not as whole water molecules, but as OH^- and H^+ ions, with the OH^- attacking the terminal phosphate of the ATP from the outside and the H^+ from the inside. This process uses up H^+ on the inside of the membrane and produces H^+ on the outside (as water dissociates to make up for the loss of OH^-), so it is as if H^+ is actually transported across the membrane by the hydrolysis of ATP.

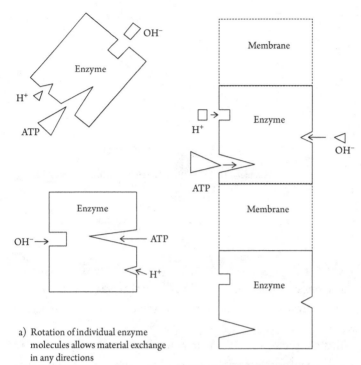

a) Rotation of individual enzyme
molecules allows material exchange
in any directions

b) Orientation and constraint of enzymes by
membrane requires ATP and H⁺ to approach
from inside and OH⁻ from outside

The result is that the hydrolysis of ATP to ADP is accompanied by the formation of a transmembrane concentration gradient of H^+, the concentration of H^+ being higher outside the membrane than inside. In the absence of any other ionic transfers across the membrane, an electrical gradient is also established, with the outside of the membrane positive to the inside. The free energy of the ATP is lost, but is compensated in part by the build-up of

ATP

$$\text{adenosine} - \text{O} - \overset{\overset{\text{O}}{\|}}{\underset{\underset{\text{OH}}{|}}{\text{P}}} - \text{O} - \overset{\overset{\text{O}}{\|}}{\underset{\underset{\text{OH}}{|}}{\text{P}}} - \text{O} - \overset{\overset{\text{O}}{\|}}{\underset{\underset{\text{OH}}{|}}{\text{P}}} - \text{OH}$$

H^+ OH^-

$$\text{adenosine} - \text{O} - \overset{\overset{\text{O}}{\|}}{\underset{\underset{\text{OH}}{|}}{\text{P}}} - \text{O} - \overset{\overset{\text{O}}{\|}}{\underset{\underset{\text{OH}}{|}}{\text{P}}} - \text{OH} + \text{HO} - \overset{\overset{\text{O}}{\|}}{\underset{\underset{\text{OH}}{|}}{\text{P}}} - \text{OH}$$

ADP   Phosphate

the electrical and concentration gradients of the H^+. By the coupling agency of the membrane, a chemical reaction is linked to a directed physical transport and the establishment of an electrical potential.

This coupling system may in some systems be as reversible as an ordinary (nonvectorial) chemical reaction. If the concentration gradient of H^+ is built up sufficiently by some other means, the H^+ gradient may drive the phosphorylation of ATP from ADP. This appears to be the case for the membranes of mitochondria and chloroplasts and for the cell membranes of prokaryotic cells (cells such as bacteria that have no nucleus). In the oxidation of glucose or other nutrients in our cells, the greatest part of ATP formation occurs in the mitochondria, where hydrogen that has been removed from the substrates is passed on to oxygen to form water. This highly energetic (potentially explosive!) reaction between hydrogen and oxygen is broken up into a number of small steps involving a coupling of chemical (electronic) reactions with H^+ transport across the membrane, so that the region outside the membrane acquires a higher H^+ concentration than the region inside. When this substrate oxidation system is combined with a reversible ADP–ATP transport system in a manner first proposed by Peter Mitchell in the 1960s (and called the *chemiosmotic coupling mechanism*), the organism has a mechanism for linking the powerful oxidation reactions to the synthesis of ATP.

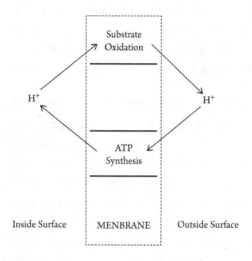

ACTIVE TRANSPORT: AN EXPERIMENTAL DEMONSTRATION

Active transport is the transmembrane movement of molecules or ions that is coupled to the energy metabolism of the cell, with the ATP-coupled transport of hydrogen ions described earlier being one example. Because active transport is coupled to an energy source, it can be driven up a concentration gradient, unlike simple diffusion, which is self-driven and must proceed down its own concentration gradient. Active transport seems to be a universal phenomenon of all living cells. Without it, the cell could not maintain its own internal composition of small molecules and ions, a composition that is very different from that of its environment. The cells of our brains and our kidneys probably use the greatest part of their oxygen and expend the largest part of their energy in maintaining active transport, especially of and K^+, across their membranes. Even the little mammalian red blood cell, which has forsaken its nucleus and mitochondria to be more fully packed with hemoglobin for carrying oxygen to other cells, retains a small glycolytic, nonoxidative metabolism, which it expends mostly on the active transport of Na^+ and K^+ across its cell membrane.

A dramatic example of active transport occurs in the skin of freshwater frogs. The body fluids of the frog are salty, containing sodium chloride (NaCl) in excess of 10^{-1} molar. The fresh water of its environment contains perhaps 10^{-5} molar NaCl. Sitting or swimming in such water, the frog would be expected to lose NaCl by diffusion across its skin because its internal concentration is 10^4 times greater than that of the solution around it. However, the Danish physiologist August Krogh showed (1937) that a frog that needed salt could actually gain it by merely sitting in very dilute salt solutions. Frog skin apparently has the ability to transport NaCl upstream, against its own concentration gradient. In the years after World War II, when radioisotopes became available for biological research, Hans Üssing (working initially in Krogh's laboratory) devised a technique for studying active transport in isolated frog skin.

When skin is removed from a frog and mounted between two chambers containing identical salt solutions, an electrical potential develops across the skin, the internal surface of the skin becoming positive to the outside surface. This electrical asymmetry, detected by electrodes placed near the skin surfaces, cannot be ascribed to

the identical salt solutions of the two chambers and must therefore
be due to a vectorial biochemistry in the skin that creates the charge
separation. The voltage difference across the skin is maintained over
time in spite of an electrical conductivity in the skin that should
passively tend to discharge the gradient, which suggests that some
other biochemical gradient is opposing discharge of the electrical
potential. This biochemical gradient necessarily involves the metabolic free energy of the skin's epithelial cells.

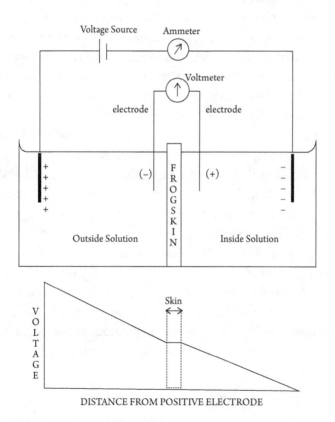

DISTANCE FROM POSITIVE ELECTRODE

The electrical potential created by the epithelial cells can be reduced by an opposing potential from a power source connected to electrodes placed outside the recording electrodes. The inside circuit with the recording electrodes contains a voltmeter for measuring potential, while the outside opposing power circuit contains an ammeter for measuring current. Fine adjustment of the opposing potential can reduce the potential across the skin to zero, with two surprising results.

First, the voltage profile from the outer positive electrode to the outer negative electrode has a discontinuity across the frog skin: the continuous voltage drop is interrupted at the skin, across which the drop is zero. It is as if the flow of positive ions from the positive to the negative electrode experiences no resistance in the skin itself. That is, the flow of ions through the skin requires no energy from the external power source because the needed energy is provided by the epithelial cells of the skin.

Second, since the potential across the skin is reduced to zero by the opposing electrodes, it may be surprising that the ammeter indicates that there *is* a current of ions through the entire system, including the skin. How can this be? To have an electrical flow, we need a gradient, but there is no purely electrical gradient across the skin. The electrical flow within the skin must be driven by an internal chemical (or electrochemical) gradient of the skin, the free energy of which comes from the metabolism of its cells. Since the flow is directional, the internal chemical gradient is vectorial, and since the flow is continuous, the gradient must be maintained by the continuous energy output of the skin. If the metabolism of the skin is blocked (e.g., by cyanide), the current largely disappears.[1]

1. As the skin's own membrane potential disappears with the inhibition of its energy metabolism, a small current is still driven across the skin by the externally imposed gradient, which now establishes its own voltage drop across the skin.

With radioactive sodium, Üssing showed that Na^+ was transported across the skin from the outside chamber to the inside and that the rate of this Na^+ transport accounted exactly for the positive current across the skin as recorded by the ammeter. The current through the skin was therefore a sodium ion current, and since this current required metabolic energy from the skin to drive it, the isolated frog skin was capable of active Na^+ transport. Üssing showed that no other ion (except lithium [Li^+] to a small degree) could substitute for Na^+ as substrate for this active ion pump. Potassium ion, however, was needed on the inside surface of the skin in order for the Na^+ pump to work. Üssing proposed that the active Na^+ pump is aligned in the epithelial cell membrane facing the interstitial fluid of the frog but not in the membrane facing the pond water. The epithelial cells do not directly pump Na^+ from the pond water, but instead pump out their own cellular Na^+ into the interstitial fluid bathing the inside layer of the skin. As the Na^+ is pumped out, it is exchanged for K^+, which enters the cell. The cell's concentration of Na^+ is lowered while its concentration of K^+ is raised. When the cell's concentration of Na^+ becomes low enough, Na^+ diffuses passively from the pond into the cell, from which it is subsequently pumped to the interstitial fluid. The K^+ brought into the cell by the Na^+ pump diffuses passively from the cell back to the interstitial fluid.

The Na^+/K^+ pump studied in frog skin by Üssing is now known to be powered by ATP and to involve an exchange of $3Na^+$ for $2K^+$ for each ATP that is hydrolyzed to ADP. The pump appears to occur in the cell membranes of all animal cells. In mammals, its action is especially notable in the cells of the nervous system; in those of heart, skeletal, and smooth muscles; in the kidneys; in the epithelial lining of the intestines; and in the red blood cells.

PHOTOSYNTHESIS: OUR BRIDGE TO THE SUN

Almost all the free energy that drives the activities of millions of living species on Earth derives ultimately from the nuclear fusion reactions going on in the interior of the Sun. Solar radiation streams across space from the intensely hot Sun to the surface of the Earth, warming all things that absorb its photons. Unequal warming, as between land and water surfaces, and especially between different latitudes, creates thermal gradients that drive our winds and ocean currents. But in the green living world of chlorophyll is a special kind of photon absorption that yields not just warmth but a transformation of the free energy of the sunlight into chemical free energy to drive the whole of the living world. This life-giving transformation is photosynthesis. Without this essential energy bridge to the Sun, almost all life on earth would cease.

In his 1845 paper "The Motions of Organisms and their Relation to Metabolism," the same paper in which he explained his 1842 calculation of the mechanical equivalence of heat, Robert Mayer saw the essential role of the green plants in the energetics of our world:

> Nature set herself the task of capturing the light flooding toward the earth, and of storing this, the most elusive of all forces, by converting it into an immobile force. To achieve this, she has covered the earth's surface with organisms which while living take up the sunlight and use its force to add continuously to a sum of chemical difference . . . . The plants take in a force, the light, and bring forth another force, the chemical difference.[2]

2. A slightly different translation from that rendered here can be found in Robert Bruce Lindsay, *Julius Robert Mayer. Prophet of Energy*. Oxford: Pergamon Press, 1973, p. 99–100. The German word *Kraft*, translated as "force" in this quotation, was used by Mayer in this context to mean what soon would be called energy.

Since prehistoric times, humanity would have been aware that light, whatever it is, interacts with matter and that matter, when heated, can give off light. Sunlight warms and makes the plants grow. Too much sunlight burns and makes the plants wilt. A burning torch gives off light, and heated metal glows. *Bioluminescence* was known to Aristotle, or to anyone before him who had seen a firefly or the foxfire of rotting wood. The Greeks of antiquity knew that sunlight could alter materials, especially when concentrated by a burning glass, which could actually start a fire. By the 17th century, the blackening of silver salts by sunlight, the basis for later film photography, had been discovered. But it was not until the discovery of the electron in 1897 by J. J. Thomson that the interactions between light and matter could be understood in modern terms. It was quickly discovered that light hitting a metallic surface could boil (so to speak) electrons off the metal, that the number of electrons fleeing from the metal was proportional to the intensity of the light, but that the velocity of the electrons (and hence their kinetic energy) depended not upon the light intensity but only upon the frequency[3] of the light and upon the particular metal employed. In general, the kinetic energy, KE, of the electrons ejected from the metal is:

$$KE = hf - w \quad \text{or} \quad hf = KE + w$$

where h is *Planck's constant*, a universal constant that applies to light of all wavelengths, f is the frequency of the light, and w is the amount of work required to separate an electron from its metal (a term that will differ for different metals). The equation expresses

3. *Frequency* is the number of vibrations of the light wave per second. Multiplied by the wavelength of the light, it gives the light velocity, c, which in a vacuum is the same for all wavelengths. As for a runner, the speed is the frequency of the strides times the length of the stride.

two important facts. The first is the conservation of energy—that the interaction of light with matter involves a free energy that is equivalent to the kinetic energy imparted to an electron plus the work needed to separate the electron from its substrate atom. The second fact portrayed by the equation is that the light can bring to the interaction only a definite quantity of energy that is linearly proportional to its frequency. Einstein saw this in 1905 as meaning that light itself, though it has properties of waves, travels not as a continuous stream but as discrete photons, each of which carries a quantum of energy equal to hf.

Most chemical reactions are not affected by light; if they were, we would have to speak of chemical reactions conducted not only at standard temperature and pressure, but at standard luminosity as well. Such reactions are thermochemical in nature: they are affected by temperature but not by light. A few chemical reactions, such as the excitation of silver in the photographic process, are greatly affected by light and hardly at all by temperature—a film camera has f-stops for controlling light, but nothing for adjusting to environmental temperature. Such reactions are photochemical reactions. Photosynthesis, however, is a complex process divisible into two parts, one of which is photochemical (called the *light reaction*) and the other thermochemical (the *dark reaction*).

In terms of materials, the overall photosynthetic process for the production of glucose is the exact opposite of the oxidative utilization of glucose:

Glucose Oxidation : $C_6H_{12}O_6 + 6O_2 \rightarrow 6CO_2 + 6H_2O$
$\Delta G^{\circ} = -686,000$ cal.

Photosynthesis : $6H_2O + 6CO_2 \rightarrow 6O_2 + C_6H_{12}O_6$
$\Delta G^{\circ} = 686,000$ cal.

In terms of free energy, they are necessarily opposite as well, so that while the oxidative reaction is down-gradient by 686,000 calories per mole of glucose, the synthetic reaction is up-gradient by the same amount. That is where the light comes in: the interactions of its photons with the electrons of the chlorophyll of the green plant (or photosynthetic microbe) supply a free energy gradient that is coupled with the uphill chemical gradients involved in the synthesis of glucose.

The overall equation for photosynthesis is deceptive, however. Because it shows equal amounts of CO_2 coming in and O_2 going out, it was assumed for at least a hundred years that the oxygen evolved comes from the carbon dioxide and that photosynthesis involves splitting the carbon dioxide into carbon and oxygen, with a subsequent hydration of the carbon to form the "carbohydrate" glucose.[4] In 1931, however, the microbiologist C. B. van Niel showed that certain photosynthetic bacteria require H_2S in order to carry out photosynthesis and that, in doing so, they produce sulfur rather than oxygen. Comparing the actions of the sulfur bacteria to green plants suggested to van Niel that since the sulfur produced by the bacteria had necessarily its origin in H_2S, it was likely that the oxygen produced by green plants came from H_2O. This extraordinary idea was soon confirmed by isotopic (tracer) experiments that showed that the O_2 evolved by green plant photosynthesis comes from the oxygen atoms of water, not those of carbon dioxide, and by the demonstration by Robin Hill that isolated chloroplasts could, when illuminated, produce O_2 even in the absence of CO_2.

---

4. The carbohydrates were so named because their basic formulae suggested that they are hydrates of carbon. For example, glucose appears to have six units of (CH_2O), while ribose has five such units and others have three, four, or seven units.

Photosynthesis thus involves the splitting of water into oxygen that is released as O_2 gas and reactive hydrogen that is used for reducing CO_2. As an electric current generated by a battery can cause water to be split into oxygen and hydrogen, so photons of light absorbed by chlorophyll can cause electron movements that split water into its elements. But while O_2 is released as gas in both cases, in photosynthesis, the hydrogen is passed in an active form to a specific biochemical carrier of hydrogen, nicotine adenine dinucleotide phosphate (NADP) to be used, in effect, for the reduction of CO_2. The photochemical (light) reaction is a vectorial one involving the membrane structure of the chloroplasts, and part of the hydrogen is ionized to H^+ to form an ion gradient across the chloroplast membranes, a free energy gradient that is used for the phosphorylation of ADP to ATP.

The capture of CO_2 by the plant does not require light. The primary reaction of CO_2, discovered through tracer experiments conducted by Melvin Calvin, is commonly with ribulose diphosphate (RDP) to form a six-carbon compound that immediately breaks down to form two three-carbon (phosphoglyceric acids; PGAs). Part of the PGA is used to form glucose and part to regenerate the RDP needed to continue the CO_2 capture. This forms a cycle of reactions known as the *Calvin cycle*, and the totality of these reactions is the dark reaction. The dark reaction is not self-sustaining, however, for it must continually be fed the ATP and NADPH (a reduced form of NADP) that are produced in the light reaction.

Through its integration of light and dark reactions, photosynthesis captures a small part of the Sun's emitted photons, which otherwise would flee without apparent purpose into outer space, and it builds from their action the biochemical gradients that have driven for billions of years the evolution of living forms, giving

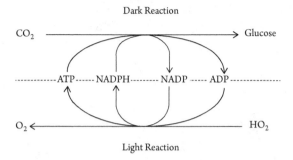

birth to consciousness and humanity and our endless quest to understand the origins and purpose of our being:

> The flowers looking in from the walled garden through my window do not, it is true, see me. But their leaves see the light, as my eyes can never do. They take it, as it forever spills away radiant into space in a golden waste, to a primal purpose. They impound its stellar energy, and with that force they make life out of the elements. They breathe upon the dust, and it is a rose.[5]

5. Donald Culross Peattie, *Flowering Earth*. New York: G. P. Putnam's Sons, 1939, p. 4

BIBLIOGRAPHY

Baldwin, Ernest. *Dynamic Aspects of Biochemistry*, 5th edition. Cambridge: Cambridge University Press, 1967. This is a highly readable introduction to the intermediary metabolism of the living organism that lies within life's overall energy relations. The author had a gift for clarity and for summarizing the experimental basis and historical development of modern biochemistry.

Baron, M. "With Clausius from Energy to Entropy." *Journal of Chemical Education, 66*, 1001–1004. 1989. This brief article outlines the development of Clausius's concept of entropy.

Bassham, J. A., and Melvin Calvin. *The Path of Carbon in Photosynthesis*. Englewood Cliffs, N.J.: Prentice-Hall, 1957. This little book relates some of the methods and results of the radioisotope studies that uncovered the sequence of biochemical reactions taking place in the dark reaction of photosynthesis

Bent, Henry A. *The Second Law*. New York: Oxford University Press, 1965. Henry Bent was one of Luke Steiner's students of thermodynamics at Oberlin College. In this admirable introduction to the laws of chemical thermodynamics, he rightfully emphasizes the need to account for the total energy changes undergone by both the system under investigation and its environment. The book contains many interesting examples and problems, with an extensive appendix of thermodynamic data.

Brazier, Mary A. B. "The Historical Development of Neurophysiology." In *Handbook of Physiology*, section 1, vol. 1, pp. 1–58. Washington, D.C.: American Physiological Society, 1959. This is a fascinating, highly detailed, and documented account of the history of neurophysiology.

Of particular interest here is the lucid account of the work of Galvani and Volta.

Brönsted, J. N. *Physical Chemistry*. Translated by R. P. Bell. New York: The Chemical Publishing Company, 1938. In his chapters on thermodynamics, Brönsted clearly distinguished between what he called *potential energy* (free energy dependent on a gradient) and what he termed *equipotential energy* (bound energy that has no gradient). Brönsted was a Danish physical chemist distinguished for his work on acid–base catalysis; this book and his later *Principles and Problems in Energetics* (see below) were translated into English by his student and colleague, R. P. Bell of Balliol College, Oxford.

Brönsted, J. N. *Principles and Problems in Energetics*. Translated by R. P. Bell. New York: Interscience Publishers, 1955. This little book deserves to be read and re-read for its unified theory of the role of gradients in natural processes. It is not easy reading, but rewards study.

Burton, Alan C. "The Properties of the Steady State Compared to those of Equilibrium as Shown in Characteristic Biological Behavior." *Journal of Cellular and Comparative Physiology 14*, 327–349. 1939. In this pioneering essay, Burton showed how simple kinetic systems can produce steady-states that mimic biological behavior in combining continual activity with constancy of composition.

Caneva, Kenneth L. *Robert Mayer and the Conservation of Energy*. Princeton: Princeton University Press, 1993. This is a scholarly biography of Robert Mayer that describes his struggle to arrive at the concept of energy and its conservation in all natural changes.

Cardwell, Donald S. L. *From Watt to Clausius. The Rise of Thermodynamics in the Early Industrial Age*. London: Heineman Educational Books, 1971. This is an immensely scholarly but beautifully written account of how the design and improvement of engines powered by natural forces of steam and water went hand in hand with the development of a science of heat, work, and energy. The works of James Watt, Sadi Carnot, and Rudolf Clausius, among many others, are lucidly described and explained.

Cardwell, Donald S. L. *Turning Points in Western Technology*. New York: Science History Publications, 1972. This is another masterpiece by Cardwell documenting the interrelationships between emerging technology and physical science from the Middle Ages through to the 20th century.

Cardwell, Donald S. L. *James Joule*. Manchester: Manchester University Press, 1989. All of Cardwell's books are fascinating, and in this one he gives an appreciative account of the life and work of the man who measured with experimental precision the quantitative relations between mechanical work, electrical activity, and heat.

Carnot, Sadi. *Reflexions on the Motive Power of Fire*. Translated and edited by Robert Fox. Manchester: Manchester University Press, 1986. Cardwell called this great classic "the most *original* work of genius in the whole history of the physical sciences and technology." It is hard to disagree with that judgment. Carnot's thought is immensely creative and a joy to read. Carnot's 1824 paper is also available in English, together with the 1834 paper of Émile Clapeyron and an 1850 memoir of Rudolf Clausius, in a volume published by Dover Publications (1960) under the editorship of E. Mendoza.

Chang, Hasok. *Inventing Temperature*. New York: Oxford University Press, 2004. How do we know that such a thing as temperature really exists? This is a fascinating historical and philosophical discussion of the invention of temperature scales. Chang is of the opinion (and I agree) that modern science has run away from its historical and philosophical foundations to such an extent that students seldom question the bases for the things they accept as true. Chang is much in the spirit of Ernst Mach (and also Socrates) in wanting to question the meaning of what we believe and how we think we know what we claim to know.

Clark, William Mansfield. *Topics in Physical Chemistry*. Baltimore: Williams and Wilkins, 1948. This is a brilliant and erudite presentation of some of the many interfaces between physical chemistry and the biological sciences. The author was especially known for his studies of pH and its effects on biological activity, but his chapters on energy in this book are illuminating and bear the honest mark of an exceptional intellect and humble spirit.

Clausius, Rudolf Julius Emmanuel. *The Mechanical Theory of Heat, with Its Applications to the Steam-Engine and to the Physical Properties of Bodies*. Translated by John Tyndall. Edited by Thomas Archer Hirst. London: John van Voorst, 1867. This is a collection of nine research memoirs in which Clausius gradually came to his formulation of entropy and determined that entropy, but not heat, is conserved in the ideal reversible working of a steam engine. This is the place to look to try to understand the evolution of Clausius's thought, which is a rewarding but not easy task. This book should not be confused with a later (1879) textbook that Clausius published with the same title of *The Mechanical Theory of Heat* (but without the subclause "with its Applications . . ."). This later book is sometimes referred to as a second edition of the earlier work, but it is not. It is an entirely different work and does not record the development of Clausius's ideas as clearly as does the 1867 book.

Collingwood, R. G. *The Idea of Nature*. Oxford: Oxford University Press, 1945. Collingwood examines the changing themes or analogies that have characterized humanity's attempt to understand nature. In Greek times, nature was often viewed as an organism, alive and mindful. In the late Middle Ages, this view was gradually supplanted by the metaphor of the machine—that nature was designed and created by God to act mechanically in the way that machines are made by people. And, in the late 18th century, this analogy began to be replaced by a historical view of nature—that nature is a progressive evolutionary process similar to that exhibited in human history. This is a truly thought-provoking survey of Western intellectual history as it relates to natural philosophy.

Coopersmith, Jennifer. *Energy the Subtle Concept*. Oxford: Oxford University Press, 2010. This is a brilliant survey of the origins and development of the concept of energy in scientific thought, written by a physicist who has examined in great detail the history of her subject. It is full of interesting perspectives and relationships as well as facts.

Cornford, Francis Macdonald. *Before and After Socrates*. Cambridge: Cambridge University Press, 1932. This little book can be read in a few hours, but it is worth an immeasurable number of credits in liberal education. It explores the origins of philosophy, including the natural philosophy we now call "science," in ancient Greece, beginning with the conception of a unified nature, of a single universe of being that can be examined by human thought. It beautifully describes the tension, still with us today, between the Pre-Socratic Ionian philosophers, who sought a constant material unity behind all natural change, and Socrates and his successors, Plato and Aristotle, who sought the purpose and goals of life and nature. Cornford was a classics scholar at Cambridge University and was married to a granddaughter of Charles Darwin.

Crosby, Alfred W. *The Measure of Reality. Quantification and Western Society, 1250–1600*. Cambridge: Cambridge University Press, 1997. How does it happen that we have come to think about nature and natural change in quantitative terms? Crosby examines with sweeping synthesis as well as penetrating analysis the rise of quantitative thinking in Western society from the middle of the 13th century until the time of Galileo and William Harvey and shows how quantitative measurement became increasingly prevalent in such diverse activities as time-keeping, commercial accounting, music, and art. Such developments set the stage for the quantitative measurements of modern science that would give us the principles of the conservation of mass and energy.

DeVoe, Howard. *Thermodynamics and Chemistry*. Upper Saddle River, N.J.: Prentice Hall, 2000. Howard DeVoe was a student of J. Arthur Campbell and Luke Steiner at Oberlin College in the 1950s, and he retained a deep interest in chemical thermodynamics in his subsequent teaching and research. This is a brilliantly clear and precise modern textbook. The author has done great service not only in writing it, but in making an updated version available free online.

Einstein, Albert. *Investigations on the Theory of the Brownian Movement*. Translated by A. D. Cowper. Edited with notes by R. Fürth. New York: Dover Publications, 1956. This deceptively little book contains five terse papers that the young Einstein wrote on the relation between Brownian motion and the kinetic theory of matter. Einstein is brief on verbal explanations and not always easy to understand, so this classic study should be recommended only to the mathematically inclined.

Fenn, John B. *Engines, Energy, and Entropy*. San Francisco: W. H. Freeman, 1982. A professor of chemical engineering at Yale, Fenn succeeded in writing a profoundly challenging yet thoroughly engaging, even entertaining, introduction to thermodynamics, with many interesting examples and problems.

Fong, Peter. *Foundations of Thermodynamics*. New York: Oxford University Press, 1963. For the theoretically inclined, this little book examines in a refreshing manner the logical and historical foundations of thermodynamics.

Fuchs, Hans U. *The Dynamics of Heat*. New York: Springer-Verlag, 1996. This is a momentous and ground-breaking work. Fuchs is a successor to Ernst Mach and J. N. Brönsted in clearly stating the relation between free energy gradients and natural change of all kinds. He presents all physical and chemical changes as flows across gradients. Unlike many authors, he clearly identifies entropy as the factor that flows across a thermal gradient.

Fuchs, Hans U. "A Surrealistic Tale of Electricity." *American Journal of Physics* 54, 907–909. 1986. In the spirit of Ernst Mach, Fuchs argues that the way we view electrical charge and electrical energy today is the result of the historical sequence in which electrical phenomena were measured.

Fuchs, Hans U. "Entropy in the Teaching of Introductory Thermodynamics." *American Journal of Physics* 55, 215–219. 1987. Fuchs addresses the question of how the concept of entropy should be approached and presented.

Gialason, Eric A., and Norman C. Craig, "Why Do Two Objects at Different Temperatures Come to a Common Temperature When Put in Contact? Entropy Is Maximized." *Journal of Chemical Education* 83 (6), 885–889. 2006. The authors have analyzed in full generality, and with clarity and precision, the equilibrium reached by two objects of different temperature placed into contact with one another. Norman Craig taught chemistry at Oberlin College for well over 40 years. He and Eric Gialason were both students of Luke Steiner and J. Arthur Campbell.

Gies, Frances, and Joseph Gies. *Cathedral, Forge, and Waterwheel. Technology and Invention in the Middle Ages*. New York: Harper Collins, 1994. This is a beautifully written survey of the technology of the Middle Ages, which saw a great increase in the harnessing of natural power to the service of everyday chores.

Gillispie, Charles Coulston. *The Edge of Objectivity. An Essay in the History of Scientific Ideas*. Princeton: Princeton University Press, 1960. If one could read only one book on the history of modern science, this might be the one to choose. It is breath-taking in its erudition and insight. Especially germane to our subject is chapter IX: "Early Energetics."

Gimpel, Jean. *The Medieval Machine. The Industrial Revolution of the Middle Ages*. New York: Holt, Rinehart, and Winston, 1976. In this fine book on the advancing technology of the Middle Ages, Gimpel views the increase in the use of water power as so striking as to deserve to be called an early industrial revolution.

Goldstein, Martin, and Inge F. Goldstein. *The Refrigerator and the Universe*. Cambridge, Mass.: Harvard University Press, 1993. The authors have done an exceptionally fine job of presenting the laws of thermodynamics to the general reader.

Guerlac, Henry. *Antoine-Laurent Lavoisier. Chemist and Revolutionary*. New York: Charles Scribner's Sons, 1975. A distinguished historian of science, Guerlac wrote extensively about Lavoisier. In this little book, he presented an engaging summary of the life and work of the man who was the foremost founder of modern quantitative chemistry.

Helmholtz, Hermann. "On the Conservation of Force." Translated by E. Atkinson. In Hermann Helmholtz, *Popular Scientific Lectures*. New York: Dover Publications, 1962. In the German scientific lexicon of the 19th century, the word "force" (*Kraft*) came to have several meanings: its Newtonian meaning as it relates to the acceleration of motion, a more general concept referring to the "cause" of a natural change, and (closely related to the former) energy as we use the term today. In this brilliant popular exposition, Helmholtz discusses the origins and meaning of the conservation of total "force" or energy.

Henderson, Lawrence J. *The Fitness of the Environment. An Enquiry into the Biological Significance of the Properties of Matter*. New York: Macmillan, 1913. Henderson was a physical chemist and physiologist at Harvard who did pioneering studies on the blood as a pH buffering and O_2 carrying system. He came to realize that the physical and chemical properties of common substances such as water and carbon dioxide are uniquely suited to the physiological requirements of life. This book is one of the classics of modern biology, showing that living and inorganic nature are united by fundamental principles of matter as well as energy, and that (turning the Darwinian tables) the environment (the entire realm of nature's laws) is fit for life in general just as life in special forms fits itself for the environment.

Huntley, H. E. *Dimensional Analysis*. New York: Dover Publications, 1967. Huntley's book is eminently readable and interesting, a fine introduction to the importance of dimensional analysis—a subject too often neglected.

Jammer, Max. *Concepts of Force. A Study in the Foundations of Dynamics*. New York: Harper Torchbooks, 1962. This is a vast and penetrating historical study of the concept of force from ancient to modern times.

Kluyver, A. J., and C. B. van Niel. *The Microbe's Contribution to Biology*. Cambridge, Mass.: Harvard University Press, 1956. See especially chapter 3 by C. B. van Niel: "Phototrophic bacteria; key to the understanding of green-plant photosynthesis."

Koenigsberger, Leo. *Hermann von Helmholtz*. Translated by Frances A. Welby. New York: Dover Publications, 1965. Koenigsberger was a mathematician and physicist and long-time friend of Helmholtz. This is an excellent and detailed account of the life and work of Helmholtz.

Laidler, Keith J. *The World of Physical Chemistry*. Oxford: Oxford University Press, 1993. This is a gold mine of highly thoughtful summaries of the history of ideas in physical chemistry, along with interesting biographical notes on the scientists themselves.

Lehninger, Albert L. *Bioenergetics*. New York: W. A. Benjamin, 1965. This is an illuminating and ground-breaking introduction to the ways in which living cells couple one chemical gradient to another in constructing the many patterns of reactions necessary for their lives. The discussion of entropy is muddy in this first edition, but is clarified somewhat in the second edition (1971).

Leicester, Henry M. "Germain Henri Hess and the Foundations of Thermochemistry." *Journal of Chemical Education*, November 1951, 581–583. A distinguished historian of chemistry, Leicester has given a clear and interesting account of Hess's pioneering studies of thermal energy in chemical reactions.

Lindley, David. *Degrees Kelvin. A Tale of Genius, Invention, and Tragedy*. Washington, D.C.: Joseph Henry Press, 2004. This is an engaging biography of William Thomson (Lord Kelvin), for whom the Kelvin (absolute) temperature scale is named. Thomson was an experimental and mathematical genius who was a central figure in the study of temperature and heat and in the development of thermodynamics (in fact, he coined the word). He sometimes made mistakes, as most good scientists do: from his studies of the rates at which bodies cool, and unaware that our planet produces heat (from radioactivity), he insisted that the Earth could not be as old as the geologists and evolutionary biologists thought it should be. He was also an inventor and an ocean sailor who took part in the laying of the trans-Atlantic telegraph cable, devising signal boosters to make the cable functional.

Lindsay, Robert Bruce. *Julius Robert Mayer Prophet of Energy*. Oxford: Pergamon Press, 1973. This is an invaluable resource for trying to understand the evolution of Robert Mayer's conviction that a constant quantity (*Kraft* or energy) is conserved throughout all natural changes or transformations. The author was a distinguished physicist and historian of physics at Brown University. He has brought together in English translation the five papers that Mayer wrote between 1841 and 1851. In addition to providing new translations of four of the papers, Lindsay has contributed a brilliant critique of Mayer's scientific contributions and a thoughtful summary of Mayer's somewhat troubled life.

Mach, Ernst. "On the Principle of the Conservation of Energy." In Ernst Mach, *Popular Scientific Lectures*. Translated by Thomas J. McCormack. LaSalle, Ill.: Open Court Press, 1986. Mach believed in getting to the historical and logical roots of scientific ideas. This popular essay on the conservation of energy rewards careful reading.

Mach, Ernst. *Principles of the Theory of Heat Historically and Critically Elucidated*. Translated by Thomas J. McCormack. Dordrecht: D. Reidel Publishing, 1986. Not for the faint-hearted or the reader in a hurry: this is a book-length study of the history and logic of our views of temperature, heat, and the laws of energy.

Magie, William Francis. *A Source Book in Physics*. Cambridge, Mass.: Harvard University Press, 1935. This is a splendid collection of original writings, many of them hard to find, from the history of physics. See especially pp. 251–255 on Robert Brown.

Mahan, Bruce H. *Elementary Chemical Thermodynamics*. New York: W. A. Benjamin, 1963. This is a beautifully organized and clearly presented introduction to chemical thermodynamics, with many worked examples and problems.

Maxwell, James Clerk. *Theory of Heat*. Mineola, N.Y.: Dover Publications, 2001. Maxwell is most famous for his highly mathematical work on electromagnetic phenomena. This lucid textbook, written for a broader audience of students, includes many topics in physics, but with an emphasis on the measurement of temperature, heat, mechanical work, and energy.

McMahon, Thomas A., and John Tyler Bonner. *On Size and Life*. New York: Scientific American Books, 1983. A physicist (McMahon) and biologist (Bonner) have combined to present this brilliant account of the importance of size to the structure and function of living organisms. Some forces or phenomena act or depend on masses, others on surfaces.

The mass-to-surface ratio increases as size increases, and this fact influences a multitude of processes in physics and biology. See especially chapter 3, "The Physics of Dimensions."

Mitchell, Peter. "Metabolism, Transport, and Morphogenesis: Which Drives Which?" *Journal of General Microbiology 29*, 25–37. 1962. In this fascinating and highly original paper, Mitchell proposed that chemical reactions directed across membranes could either produce transport in space or be driven by it, depending on the circumstances.

Mitchell, Peter. "Chemiosmotic Coupling in Oxidative and Photosynthetic Phosphorylation." *Biological Reviews 41*, 445–502. 1966. Mitchell received the Nobel Prize for his work demonstrating that mitochondria and chloroplasts use vectorial chemistry to build gradients that energize the formation of ATP.

Mysels, Karol J. *Introduction to Colloid Chemistry*. New York: Interscience Publishers, 1959. For topics dealing with the physics and physical chemistry of macromolecules, this book is one of my favorites. It is exceptionally clear and precise. See especially chapter V: "Diffusion and Brownian Motion."

Nye, Mary Jo. *Molecular Reality. A Perspective on the Scientific World of Jean Perrin*. London: Macdonald, 1972. In the late 19th century, not all physicists and chemists believed in the reality of discrete atoms and molecules. Ernst Mach was one of the doubters. In a scholarly but eminently readable account, Nye surveys this debate, with special emphasis on the work of Jean Perrin who, by his experimental studies of Brownian motion, convinced most of the doubters (but not Mach) of the reality of discrete molecules.

Pacey, Arnold. *The Maze of Ingenuity. Ideas and Idealism in the Development of Technology*. London: Allen Lane, 1974. (Reprinted by MIT Press, 1976, with a new preface by the author.) Pacey has written more than just a masterful account of the development of technology during the second millennium AD. His is a book that bridges social, economic, and technological history with religion, philosophy, and science, showing that practical, intellectual, and spiritual motivations are intertwined in the rise of technology and science.

Park, David. *The Images of Eternity. Roots of Time in the Physical World*. Amherst, Mass.: University of Massachusetts Press, 1980. Directional time is an integral part of the second law of thermodynamics, of the principle that natural changes are caused by gradients that are diminished by the changes they cause. Yet this sort of time is not evident in Newton's universe or in his laws of mechanics. David Park, physicist and historian of science, ponders the deep philosophical puzzle of time.

Perrin, M. Jean. *Brownian Movement and Molecular Reality*. Translated by F. Soddy from *Annals de Chimie et de Physique*, September 1909. London: Taylor and Francis, 1910. Reprinted in Mary Jo Nye, *The Question of the Atom*. Los Angeles: Tomash Publishers, 1984, pp. 505–601. This is Perrin's classic experimental study of the quantitative aspects of Brownian motion, which he related to the existence of discrete molecules and the kinetic theory of matter.

Perrin, Jean. *Atoms*. Translated by D. L. Hammick from the French edition of 1913. Woodbridge, Conn.: Ox Bow Press, 1990. When this book was written, the electron, radioactivity, and radioactive transmutation were recent discoveries. Perrin presents the case for the belief in the existence of discrete atoms in chemistry and physics. Of particular interest are chapters 2–5, which discuss Brownian motion and the kinetic theory of matter.

Schoenheimer, Rudolph. *The Dynamic State of Body Constituents*. New York: Hafner, 1964. (Reprint of edition published by Harvard University Press, 1942.) The open steady-state condition of the animal body is dramatically on display in this classic biochemical study employing isotopic tracers.

Spanner, D. C. *Introduction to Thermodynamics*. London: Academic Press, 1964. Written by a plant physiologist-biophysicist, this is an engaging introduction to thermodynamics and its applications to such topics in biology as osmosis, the Donnan equilibrium, electrical potentials, and other membrane phenomena. It is a highly intelligent and attractive book—and not just because each chapter is headed by a quotation from Winnie the Pooh.

Steiner, Luke E. *Introduction to Chemical Thermodynamics*. New York: McGraw-Hill, 1941. Something about the way that Luke Steiner taught chemistry conveyed to his students a deeply probing mind. He wanted us to join him in trying to get to the fundamentals of things. That spirit is evident, too, in his fine textbook on chemical thermodynamics. It still rewards study today.

Strong, Laurence E., and Wilmer J. Stratton. *Chemical Energy*. New York: Reinhold, 1965. Strong and Stratton were professors of chemistry at Earlham College. Their book is beautifully organized and presented, and while the discussion of free energy seems at times somewhat inconsistent to me, this is an excellent general introduction to the energy of chemical reactions.

Üssing, H. H. "Transport of Ions Across Cellular Membranes." *Physiological Reviews 29*, 127–155. 1949. Üssing showed that frog skin can transport ions up a concentration gradient when the process is coupled to other down-gradient metabolic reactions. In this review essay, he surveys the studies of active cell membrane transport that were being conducted in the years immediately after World War II, when radioisotopes first became widely available for biological research.

Van Ness, H. C. *Understanding Thermodynamics*. New York: McGraw-Hill, 1969. Reprinted by Dover Publications, 1983. This is a brief but clearly presented account of the meaning of the first and second laws of thermodynamics.

Von Bertalanffy, Ludwig. "The Theory of Open Systems in Physics and Biology." *Science 111*, 23–29, 1950. Von Bertalanffy's writings are invariably highly creative and thought-provoking. In this lucid paper, the characteristics of open systems in nature are discussed. In many ways this paper is a sequel to Alan Burton's paper on steady-states.

Waser, Jürg. *Basic Chemical Thermodynamics*. Menlo Park, Calif.: W. A. Benjamin, 1966. Similar to Mahan's book, but a little more detailed, this is another fine introduction to chemical thermodynamics, with many quantitative examples and problems.

Westfall, R. S. *Force in Newton's Physics*. London: Macdonald, 1971. This is a magnificent historical study of the physics of Galileo, Descartes, Huygens, Leibniz, and Newton.

Wheeler, Lynde Phelps. *Josiah Willard Gibbs. The History of a Great Mind,* revised edition. New Haven, Conn.: Yale University Press, 1952. This is an interesting biography of one of the greatest American scientists of the 19th century, whose papers on thermodynamics were highly valued by James Clerk Maxwell and other physicists abroad.

Wightman, William P. D. *The Growth of Scientific Ideas*. Edinburgh: Oliver and Boyd, 1959. This is a highly readable topical history of science. See especially chapter XXIII, "The 'Go' of Things," which discusses the origins of thermodynamics.

Williams, L. Pearce. *Michael Faraday*. New York: Basic Books, 1965. Every page of this scholarly biography of Faraday is interesting, rich with details of his life and personality and thorough in its discussion of his experiments and ideas.

Zemansky, Mark W. *Heat and Thermodynamics*. New York: McGraw-Hill, 1957. A classic textbook, this book remains a highly useful reference.

INDEX

absolute temperature, 79, 102–6, 113, 114

absolute zero, 112, 113

acceleration, 44, 46, 47, 55–57

acetaldehyde, 175

acetic acid, 175

active transport, 235–39

adenosine diphosphate (ADP), 187, 190, 219–22, 225, 226, 232–35, 239, 244

adenosine triphosphate (ATP), 187, 190, 219–23, 224, 225, 226–27, 244

  diffusion and, 164

  free energy and, 80

  sodium/potassium pump and, 239

  transport system, 232–35

adiabatic compression, 135, 144–45, 146

adiabatic expansion, 132, 134, 144–46

ADP. *See* adenosine diphosphate

angles, 51, 52, 57–58

Archimedes, 44

area, 47

Aristotle, 17, 23, 28, 150, 169, 241

atomic-kinetic theory of matter, 150–51

atoms, 168–69

ATP. *See* adenosine triphosphate

ATPase, 232

attractive forces, 169

Avogadro, Amedeo, 2

Avogardro number, 150

balance of nature, 212

Balliol College, 3, 4

batteries, 19, 197

  free energy and, 73–74

  Joule's work and, 33–36, 62

  Volta's invention of, 20

Becquerel, Henri, 40, 64

Bell, R. P., 4

Berthelot, Marcellin, 168

Bioenergetics (Lehninger), 110

biological energy, 207–45

  active transport and, 235–39

  coupled reactions and increasing order, 216–30

  coupled reactions with transport, 231–35

  open steady-state free energy utilization, 214–16

  open systems, sustained but stable activity in, 207–14

  photosynthesis and, 212–14, 240–45

bioluminescence, 241

Black, Joseph, 24, 114, 131

blood circulation, 152

bloodletting, 25–27

Gay-Lussac, Joseph-Lewis, 2
Gibbs, Josiah Willard, 99, 110, 206
Gibbs free energy, 127n2, 178–87
  of a chemical reaction, 180–87
  defined, 178
  of formation of a substance,
    179–80, 181–83
Glasgow University, 131
glucose, 187, 190, 217–19, 220, 222, 234,
  242–43, 244
glucose-1-phosphate, 223–24,
  225, 228
glucose-6-phosphate, 187, 190,
  222–24, 227
glycogen, 225, 226
gradients
  biological energy and, 209, 211, 212,
    216, 217, 224, 226, 238
  chemical, 165
  concentration (see concentration
    gradients)
  currents moving across, change
    as, 60–62
  down-, 218
  electrical, 172 (see also membrane
    potentials)
  free energy in equilibria of
    opposing, 81–89
  gravitational (see sedimentation
    equilibrium)
  pressure (see osmotic pressure)
  temperature, 92–93, 133–37
  up-, 218
Gravesande, Willems, 14
gravity, 11, 44, 51, 70–71
Great Pyramid at Gizeh, 43–44
Greece, ancient, 6, 25, 41, 150, 241
gun, firing of, 14–15
gunpowder, 168

Haemastaticks (Hales), 26
Hales, Stephen, 25–26
Halley, Edmund, 98
Harvey, William, 8, 25, 28, 129
heat. See also entropy; thermal energy
  as bound energy, 73–74

equivalence/transformational content
  of, 108
as free energy, 92
internal energy and, 170–75
latent, 24, 114, 131
law of constant heat
  summation, 167
as motion, 92
standard unit of (see calories)
as a substance,
  problems with, 91–92
heat capacity, 113–15, 176–77
heat engines, 140–46. See also steam
  engines
heat pumps, 140–46
heat-mechanical work relationship, 23–40,
  91, 92, 147
  Brownian motion and, 162–63
  early work on, 24–40 (see also
    Helmholtz, Hermann; Joule, James
    Prescott; Mayer, Robert)
  explained, 23–24
height, 11, 16, 34
Helmholtz, Hermann, 7, 8–9, 19,
  24, 65, 72, 74, 135, 147, 166,
  206, 207
  Kant's influence on, 18, 36
  overview of work, 36–40
  Ueber die Erhaltung der Kraft, 38
Helmholtz free energy, 127n2, 178
Henderson, Lawrence J., 228
Henry's law, 79
Heracleitus, 210
Hess, Germain Henri, 166, 206
Hess's law, 166–68, 175
heterogeneity, 152, 154–57
Hevesy, George, 121–22, 190
Hill, Robin, 243
History of the Warfare of Science and
  Theology, A (White), 9
homogeneity, 55, 56, 157. See also isolated
  homogeneous objects
Hume, David, 18
humors of the body, 25, 96
Huygens, Christiaan, 13, 130
hydrocarbons, 167